DEEP MAPS AND SPATIAL NARRATIVES

THE SPATIAL HUMANITIES
David J. Bodenhamer, John Corrigan, and Trevor M. Harris, editors

Geographies of the Holocaust
Edited by Anne Kelly Knowles, Tim Cole, and Alberto Giordano

Locating the Moving Image: New Approaches to Film and Place
Edited by Julia Hallam and Les Roberts

The Spatial Humanities: GIS and the Future of Humanities Scholarship
Edited by David J. Bodenhamer, John Corrigan, and Trevor M. Harris

Toward Spatial Humanities: Historical GIS and Spatial History
Edited by Ian N. Gregory and Alistair Geddes

Troubled Geographies: A Spatial History of Religion and Society in Ireland
Ian N. Gregory, Niall A. Cunningham, C. D. Lloyd,
Ian G. Shuttleworth, and Paul S. Ell

DEEP MAPS
and
SPATIAL NARRATIVES

EDITED BY

DAVID J. BODENHAMER,

JOHN CORRIGAN,

and

TREVOR M. HARRIS

INDIANA UNIVERSITY PRESS   *Bloomington & Indianapolis*

This book is a publication of

INDIANA UNIVERSITY PRESS
Office of Scholarly Publishing
Herman B. Wells Library 350
1320 East 10th Street
Bloomington, Indiana 47405 USA

iupress.indiana.edu

Telephone 800-842-6796
Fax 812-855-7931

© 2015 by Indiana University Press
All rights reserved

No part of this book may be reproduced or utilized in any form or by any means, electronic or mechanical, including photocopying and recording, or by any information storage and retrieval system, without permission in writing from the publisher. The Association of American University Presses' Resolution on Permissions constitutes the only exception to this prohibition.

∞ The paper used in this publication meets the minimum requirements of the American National Standard for Information Sciences—Permanence of Paper for Printed Library Materials, ANSI Z39.48–1992.

Manufactured in the United States of America

Cataloging information is available from the Library of Congress.

ISBN 978-0-253-01555-6 (cloth)
ISBN 978-0-253-01560-0 (paperback)
ISBN 978-0-253-01567-9 (ebook)

1 2 3 4 5 20 19 18 17 16 15

CONTENTS

Acknowledgments vii

Introduction: Deep Maps and the Spatial Humanities 1

1 Narrating Space and Place / *David J. Bodenhamer* 7

2 Deep Geography—Deep Mapping: Spatial Storytelling and a Sense of Place / *Trevor M. Harris* 28

3 Genealogies of Emplacement / *John Corrigan* 54

4 Inscribing the Past: Depth as Narrative in Historical Spacetime / *Philip J. Ethington and Nobuko Toyosawa* 72

5 Quelling Imperious Urges: Deep Emotional Mappings and the Ethnopoetics of Space / *Stuart C. Aitken* 102

6 Deep Mapping and Neogeography / *Barney Warf* 134

7 Spatializing and Analyzing Digital Texts: Corpora, GIS, and Places / *Ian Gregory, David Cooper, Andrew Hardie, and Paul Rayson* 150

8 GIS as a Narrative Generation Platform / *May Yuan, John McIntosh, and Grant DeLozier* 179

9 Warp and Weft on the Loom of Lat/Long / *Worthy Martin* 203

Conclusion: Engaging Deep Maps 223

Contributors 235
Index 239

ACKNOWLEDGMENTS

The editors gratefully acknowledge the support of the National Endowment for the Humanities and the Indiana University–Purdue University Indianapolis Arts and Humanities Institute in the preparation of this work. Any views, findings, conclusions, or recommendations expressed in this publication do not necessarily reflect those of the National Endowment for the Humanities or the Indiana University–Purdue University Indianapolis Arts and Humanities Institute.

We thank Robert Sloan for his support and collaboration, Michelle Sybert for her expert management of production, and Charlie McGrary for his great work on the index. We are especially grateful for the careful reading and excellent suggestions made by the two readers enlisted by IU Press.

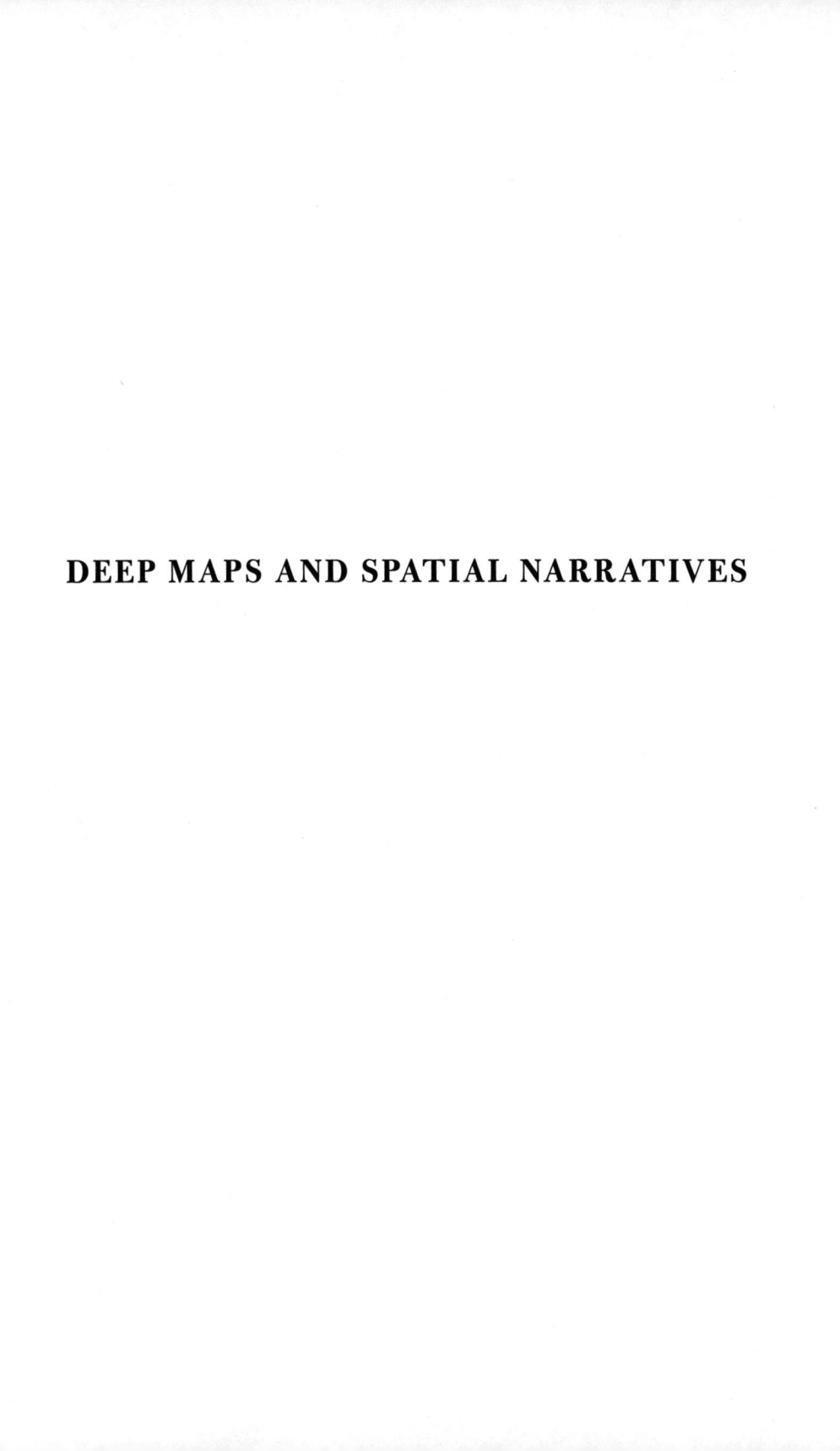

DEEP MAPS AND SPATIAL NARRATIVES

INTRODUCTION

Deep Maps and the Spatial Humanities

The word *deep* has become academic kudzu, a wildly proliferating adjective that attaches itself onto everyday concepts and often makes them impenetrable to average readers. Consider the following examples:

- Deep learning: a subfield of machine learning that is based on learning several levels of representations, corresponding to a hierarchy of features or factors or concepts, where higher-level concepts are defined from lower-level ones, and the same lower-level concepts can help to define many higher-level concepts.
- Deep processing: memory-formation involving elaboration rehearsal which involves a more meaningful analysis (e.g., images, thinking, associations) of information and leads to better recall.
- Deep structure: a theoretical construct in linguistics that seeks to unify several related structures.

Deep mapping adds to this list, not from any desire to make obscure what seems plain but rather because it is the essential next step for humanists who are eager to take full advantage of the spatial turn that already has begun to shape our disciplines.

Humanities scholars are becoming increasingly aware of the importance of geographic information. We can point to a number of causes for this development—the emergence and rapid maturation of geographic information systems (GIS) as a core technology, the convergence of web and mobile technologies that moved spatial data and its manipulation

1

beyond the realm of specialist tools, and the explosive growth of a global economy with its demand for location-based information. We also have discovered that spatially oriented software, represented by GIS, facilitates the integration of data that is so essential to our paradigmatic shift toward interdisciplinary research. We have been reminded as well of the power of the map to display information cartographically in a manner that provides fresh perspective and new insights into the study of culture and society. For all these reasons and more, we stand at the threshold of what promises to be a new age of discovery in the humanities.

The spatial humanities are being profoundly influenced by these developments. At first glance, this argument may seem odd. It runs counter to recent critiques that GIS rests on a positivist epistemology and demands a precision in data and methods much more suited to the social sciences than to the humanities. GIS also has difficulty handling time, the sine qua non for most humanities disciplines. But increasingly spatial technologies are being used in tandem with web applications in ways that make them eminently suitable for humanities scholarship, and it is this combination that promises a revolution in the ways we think about the past.

Humanists view the world as extremely complex, with endless connections among events and actors and multiple causes for effects that exert continuing influence on the world of thought and behavior. This sense of weblike interrelatedness plays itself out within two dimensions—space and time. Although the past is always bound by these two elements, humanists often treat them as artificial, malleable constructs. We move freely across these spatial and temporal grids, ignoring issues of scale, as we compare and contrast one place or one time with another in an effort to recapture a sense of the whole, to illuminate differences, and to discover patterns.[1] For the humanist, space is not only physical space but occupied space, or place, and the concept, like that of time, exists not simply in a real world but in memory, imagination, and experience. Such casual use of time and space is a curious circumstance for a discipline that, in so many ways, refers to these terms continually. An explanation lies in recognizing the ends of scholarship: the historian, for example, seeks to simulate a world that is lost, not to recreate it precisely or use it for predictive purposes. Traditionally, historians have used narrative to construct the portrait that furthers this objective. Narrative encourages the interweaving of evidentiary threads and permits the scholar to qualify, highlight, or subdue any thread or set of them. It uses emphasis, nuance, and other literary devices to achieve the complex construction of culture, past and present.

Trying to comprehend space, place, and time in concert has always proven difficult, even in the most expert narratives. Historian Hugh Trevor-Roper noted the problem decades ago: "How can one both move and carry along with one the fermenting depths which are also, at every point, influenced by the pressure of events around them? And how can one possibly do this so that the result is readable?"[2] Or as digital humanities pioneer Edward Ayers has asked more recently, "how might we combine the obvious strengths of geographic understanding with the focus on the ineffable, the irreducible, and the particular . . . ? How might we integrate structure, process, and event? In sum, how might we combine space, time, and place?"[3]

It is here where the deep map becomes important, perhaps essential. A deep map is a finely detailed, multimedia depiction of a place and the people, animals, and objects that exist within it and are thus inseparable from the contours and rhythms of everyday life. Deep maps are not confined to the tangible or material, but include the discursive and ideological dimensions of place, the dreams, hopes, and fears of residents—they are, in short, positioned between matter and meaning. They are also topological and relational, revealing the ties that places have with each other and tracing their embeddedness in networks that span scales and range from the local to the global. The spatial considerations remain the same, which is to say that geographic location, boundary, and landscape remain crucial. What is added by these deep maps is a reflexivity that acknowledges how engaged human agents build spatially framed identities and aspirations out of imagination and memory and how the multiple perspectives constitute a spatial narrative that complements the prose narrative traditionally employed by humanists.

A deep map is simultaneously a platform, a process, and a product. It is an environment embedded with tools to bring data into an explicit and direct relationship with space and time; it is a way to engage evidence within its spatiotemporal context and to trace paths of discovery that lead to a spatial narrative and ultimately a spatial argument; and it is the way we make visual the results of our spatially contingent inquiry and argument. Within a deep map, we can develop the event streams that permit us to see the confluence of actions and evidence; we can use path markers or version trackers to allow us (and others) to trace our explorations; and we can contribute new information that strengthens or subverts our argument, which is the goal of any exploration. It is, in short, a new creative space that is visual, structurally open, genuinely multimedia and multilayered. Deep maps do not explicitly seek authority or objectivity but provoke negotiation between insiders and outsiders, experts and contributors, over what is represented and how. Framed as a conversation and not a statement, they are inherently unstable, continually unfolding and changing in response to new data, new perspectives, and new insights.

The essays in this book investigate deep mapping and the spatial narratives that stem from it. They were first delivered in an expert workshop held in Denver in March 2012; the aim was to develop a theoretical and practical understanding of deep maps and spatial narratives for an NEH Advanced Institute by that name which took place in Indianapolis over the last two weeks of June 2012.[4] The authors come from a variety of disciplines, as befits the nature of the challenge: history, religious studies, geography and geographic information science, and computer science. Each one has an established reputation in creatively applying the concepts of space, time, and place to problems central to an understanding of society and culture.

Revised in response to critiques at the workshop and institute, these essays set forth a framework for understanding and constructing deep maps and spatial narratives, as well as for evaluating the promise they hold for the spatial humanities. Stuart Aitken and Barney Warf, both geographers, identify how deep maps work in practice, with Aitken focusing on the power of emotional mapping and Warf exploring how the deep map is an open framework for bridging expert and native knowledge. John Corrigan and David Bodenhamer, religious studies and history, respectively, raise questions about the process of emplacement and narrative structure within a deep mapping environment. GIS cientist May Yuan and historical geographer Ian Gregory discuss how humanists can mine text effectively for the purposes of deep mapping, Yuan through computer-aided parsing and Gregory through a marriage of computational linguistics and GIS.

Another GIS scientist, Trevor Harris, finds great value in the deep map as an intensive multimedia and immersive environment. Worthy Martin, a computer scientist, uses examples of early efforts at deep mapping to understand how it can create more nuanced understandings of human experience, while historians Phil Ethington and Nobuko Toyosawa provide a rich example of a ghost map in which depth is achieved by keeping the past visible in the present.

Through these essays, we can begin to grasp the potential of deep maps and spatial narratives in at least two ways. On one path, these web-based spatial technologies offer a powerful framework for managing and analyzing evidence, contributing primarily by locating historical and cultural exegesis more explicitly in space and time. They aid but do not replace the traditions developed over centuries by humanists: they find patterns, facilitate comparisons, enhance perspective, and illustrate data, among other benefits, but the results ultimately find expression primarily in the vetted forms accepted by our disciplines. On a second path, the deep map offers the potential for an open, unique postmodern scholarship that embraces multiplicity, simultaneity, complexity, and subjectivity. In it, we do not find the grand narrative but rather a spatially facilitated understanding of society and culture embodied by a fragmented, provisional, and contingent argument with multiple voices and multiple stories. The deep map offers a way to integrate these multiple voices, views, and memories, allowing them to be seen and examined at various scales. It will create the simultaneous context that we accept as real but unobtainable by words alone. By reducing the distance between the observer and the observed, it promises an alternate view of history and culture through the dynamic representation of memory and place, a view that is visual and experiential, fusing qualitative and quantitative data within real and conceptual space.

Above all, we offer these essays as an invitation to consider which path (or paths) we should take.

DB, JC, and TH
October 2013

NOTES

1. John Lewis Gaddis, *The Landscape of History: How Historians Map the Past* (New York: Oxford University Press, 2002), 53–71.

2. Quoted in K. Thomas, "A Highly Paradoxical Historian," *New York Review of Books*, April 12, 2007, 53–57.

3. Edward L. Ayers, "Turning toward Place, Space, and Time," in David J. Bodenhamer, John Corrigan, and Trevor M. Harris, eds., *The Spatial Humanities: GIS and the Future of Humanities Scholarship* (Bloomington: Indiana University Press, 2010), 1–13.

4. For more on the work of the NEH Advanced Institute, see "Spatial Narratives and Deep Maps: A Special Report," *The International Journal of Humanities and Arts Computing* 7, no. 1–2 (2013), 170–227.

1

NARRATING SPACE AND PLACE

DAVID J. BODENHAMER

Over the past two decades, the humanities and social sciences especially have advanced a more complex and nuanced understanding of space. For nongeographers, this intellectual movement, often labeled the "spatial turn," has been largely defined by a greater awareness of place, manifested in specific sites where human action occurs. Subject matter once organized largely by periods of time, with names such as the Great Depression or the Age of Discovery, now embraces themes of region, diaspora, contact zones, and borders or boundaries. Interest in the material and cultural markers of space and place has reinforced this shift. As a result, our sense of space and place has become more complex and problematic, but in the process it has assumed a more interesting and active role in how we understand history and culture.[1]

It is not the first time that attention to space and time has reshaped the way we approach social and cultural questions. A similar turn occurred from 1880 to 1920 when distance-collapsing innovations—the telephone, wireless telegraph, radio, cinema, automobiles, and airplanes, among others—challenged traditional understandings of how time and space intersected with the social world. It suddenly was possible to know events as they happened, and this experience of simultaneity refashioned people's sense of distance and direction. It also meant that individuals were no longer cut off from the flow of time; widely available film and photographic images made the past as accessible as the present, while new developments in science and the World Fairs that showcased them made the future seem more definite and real. New scientific theories, business practices, and cultural forms reinforced the shift: Einstein's theory of relativity and Freud's conception of psychoanalysis shaped consciousness directly; time-management studies, known as Taylorism, dominated manufacturing; and James Joyce and Marcel Proust explored how to link time and space in novels, while the Cubists challenged notions of spatial perspective and form that had long dominated art.[2]

A continuous thread links the first spatial turn with the one we have experienced more recently, but it is likely that this second turn will have a more profound influence on the theory and practice of the humanities, in large measure because of the digital revolution that has accompanied and facilitated it. The early twentieth-century reworking of space and time had less effect on the study of heritage and culture than it did on the study of art and literature. The frontier thesis of Frederick Jackson Turner and the emphasis on the American western history it spawned were exceptions, as was the decades-long work of the *Annales* school of historiography; both reflected an intentional focus on spatial questions. But the cataclysms of the mid-twentieth century, from world wars and revolutions to mass movements for equality, spurred historians to search for the roots of momentous events in ideas and politics and technological or social change, causes for which spatial markers were less pronounced. The consideration of space did not disappear, but it became marked by particularity, an emphasis on place, as scholars began to discern how the story of change differed from one location to another. This focus on the local reflected and reinforced a postmodernist unease with the grand narrative, which created a literature that increasingly became fragmented, with analyses existing at different geographical and temporal scales and few efforts made to link them. For many humanists, space itself became less geographical, as scholars found richer meaning in conceptual space—for instance, gendered space, racialized space, or the body as space—than in categories related to the physical environment, the traditional frame of definition for spatial terms.[3]

Today, historians and other humanists are acutely aware of the social and political construction of space and its unique expression as place. Space is not simply the setting for historical action but is a significant product and determinant of change. It is not a passive setting but the medium for the development of culture: "space is not an empty dimension along which social groupings become structured," sociologist Anthony Giddens notes, "but has to be considered in terms of its involvement in the constitution of systems of interaction."[4] All spaces contain embedded stories based on what has happened there. These stories are both individual and collective, and each of them link geography (space) and history (time). More important, they all reflect the values and cultural codes present in the various political and social arrangements that provide structure to society. In this sense, then, the meaning of space, especially as place or landscape, is always being constructed through the various contests that occur over power. There is nothing new in this development—the earliest maps reveal the power arrangements of past societies—but humanities scholarship increasingly reflects what may in fact be the greatest legacy of postmodernism, the acknowledgment that our understanding of the world itself is socially constructed.[5]

At its core, the current spatial turn rejects the universal truths, grand narratives, and structural explanations that dominated the social sciences and the humanities during much of last century. Above all, it is about the particular and the local, without any supposition that one form of culture is better than another. Its claim is straightforward: to understand human society and culture we must understand how it developed in certain circumstances and in certain times and at certain places. From this knowledge, we can appreciate that the world is not flat but incredibly complicated and diverse. This view no longer seems new because humanists have embraced it eagerly; now, we all recognize the particularity of space, the importance of place. But for all the uses we make of this insight—and for all its explanatory power—the concepts of space and place employed by humanists frequently are metaphorical and not geographical. Far less often have we grappled with how the physical world has shaped us or how in turn we have shaped perceptions of our material environment.

New spatial technologies, especially geographic information systems (GIS), are aiding this rediscovery of physical space in the humanities. Within a GIS, users can discern relationships that make a complex world more immediately understandable by visually detecting spatial patterns that remain hidden in texts and tables. Maps have served this function for a long time, but GIS brings impressive computing power to this task. Its core strength is an ability to integrate, analyze, and make visual a vast array of data from different formats, all by virtue of their shared geography. This capability has attracted considerable interest from historians, archaeologists, linguists, students of material culture, and others who are interested in place, the dense coil of memory, artifact, and experience that exists in a particular space, as well as in the coincidence and movements of people, goods, and ideas that have occurred across time in spaces large and small. Recent years have witnessed a wide-ranging, if still limited, application of GIS to historical and cultural questions: Did the Dust Bowl of the 1920s and 1930s result from overfarming the land or was it primarily the consequence of larger-term environmental changes? What influence did the rapidly changing cityscape of London have on literature in Elizabethan England? What is the relationship between rulers and territory in the checkered political landscape of state formation in nineteenth-century Germany? How did spatial networks influence the administrative geography of medieval China? What spatial influences shaped the development of the transcontinental railroad network in the United States? Increasingly, scholars have turned to GIS to provide new perspective on these and other topics that previously have been studied outside of an explicitly spatial framework.[6]

Despite this flurry of interest and activity, the humanities pose epistemological and ontological issues that challenge the technology in a number of ways, from the imprecision and uncertainty of humanities data to humanists' reliance on time (and time linked to space) as an organizing principle. Essentially, GIS and its related technologies currently allow users to determine a geometry of space; fuzzy data, conceptual space, and relative time pose often-insurmountable problems for these tools. In the context of the humanities, it will be necessary to replace this more limited quantitative representation of space with a view that emphasizes the intangible and socially constructed world and not simply the world that can be measured. It also will be essential to match technologies with the traditions of argument and narrative employed by humanists.

The goal of humanities scholarship is not to model or replicate the past but rather to pursue the fullest possible understanding of heritage and culture. Questions drive historical scholarship, not hypotheses, and the questions that matter most address causation: why matters more than whom, what, or when, even though these latter questions are neither trivial nor easy to answer. The research goal is not to eliminate explanations or to disprove the hypothesis but to open the inquiry through whatever means are available and by whatever evidence may be found. This sense of eclectic borrowing has long informed humanities scholarship and even finds strong advocates among some of the most well-known theorists in the humanities, hence the advice offered by Paul de Man to develop "a new kind of skill... the capacity to use and feel at home in a whole series of different critical and theoretical codes and systems, as one would use a particular foreign language, without remaining rigidly locked into any one of them, but rather developing the capacity to translate those findings into different codes, systems, critical positions, as the case may require."[7]

Humanists develop knowledge through the process of argument, but a well-presented argument often does not settle a question; it may complicate it or open new questions that previously were unimagined. Similarly, humanists are hard-pressed to identify a preferred method because each avenue of investigation yields different evidence and thus different insights. Their approach is recursive, not linear: the goal is not so much to eliminate answers as to admit new perspectives. These methods doubtless appear quixotic to nonhumanists because they do not lead to finality. But for humanists, the goal is not proof but meaning. The goal of scholarship is less to produce an authoritative or ultimate answer than to prompt new questions, develop new perspectives, and advance new arguments or interpretations.[8]

Traditionally, humanities scholars have used narrative to construct the portrait that furthers this objective. Narrative encourages the interweaving of evidentiary threads and permits the scholar to qualify, highlight, or subdue any thread or set of them—to use emphasis, nuance, and other literary devices to achieve the complex construction of past or present worlds. An ancient device common to most cultures, narrative allows us to make sense of complexity, to find structure and meaning in the otherwise chaotic rush of events that occur in all times and places. Historical narratives, the foremost form among humanities disciplines, share much in common with fictional narratives, except for their claim to objectivity and their reference to evidence or works beyond the narrative itself.

In each narrative form, historical or fictional, the story is told, not lived, with textual unity imposed by the storyteller, not by history or culture itself.[9] And like their fictional counterparts, they require a sympathetic imagination that allows us "to see small and think big," as literary critic Adam Gopnik terms it, as well as to disassemble the big nouns into the small acts that make them up.[10]

But what is a narrative? It is not merely a sequence of events arranged chronologically. A list of this sort is a chronicle, which we recognize easily but have difficulty comprehending. Our nature requires that we seek meaning in events; we search for connections among events that will lead to something we can use to solve a problem, to understand an enigma, or simply to frame what we know. The story, a plot, serves as this framing device; it allows us to evaluate the relative significance of events and to relate them to whatever concerns us. It is the plot that marks the shift from chronicle to narrative: its classical form, as Aristotle noted, "has beginning, middle, and end"—and the "end is everywhere the chief thing" because it provides unity and meaning, thereby allowing us to judge an action by its results.[11]

The Western style of narrative emerged over five thousand years ago from an oral tradition and maintained many of its characteristics for centuries. Its classic expression was the epic, a category that contains a number of forms—sacred myths, legends, and folktales, among them—but its central purpose was to recreate the traditional story. The narrator's obligation was not to fact or entertainment, but to the *mythos* itself (*mythos* is the ancient Greek word for the story or plot). "Knowing, and thus being able to tell," film theorist Edward Branigan reminds us, "is a fundamental property of narration."[12] Over time, this narrative form evolved in two antithetical directions—the empirical and the fictional, the first marked by an emphasis on accurate measurement in time and space and on causality and the second by its allegiance to an ideal, truth, or beauty, instead of fact. By the seventeenth and eighteenth centuries, a new synthesis had emerged in the novel, which sought to reconcile these two threads. Since then, the novel has become the dominant form of literature and has had a powerful influence on the way humanists use narrative to conceptualize a story, illustrate reality, provide perspective, and develop meaning.[13] It also has shaped the way we represent space.

From the time of Kant, Western society has regarded time and space as the two fundamental categories that structure human experience. Narrative theory privileges the former by emphasizing the sequence of events—"Narrative gives us what may be called the shape of time"[14]—and all narratives imply a world of spatial extension. Some theorists, however, recognize an explicit linkage of time and space, such as is reflected in Mikhail Bakhtin's concept of the chronotope (time space) as the "intrinsic connectedness of temporal and spatial relationships," with time supplying the fourth dimension of space.[15] In this view, narrative is the "representation of movement within the coordinates of space and time," with events marked by the intersection of horizontal and vertical axes in a dynamic interplay between surface and depth.[16]

Narrative space, however, is not a simple construct, a mere representation of the world (fictional or real) that acts as a container for events; instead, it encompasses several types of spaces, all of which have implications for spatial narratives. It is the setting or the physically existing environment in which actions occur and through which people move, but even this concept involves more than we might assume intuitively. This spatial setting, or spatial world, of the narrative unfolds temporally as readers progress through the text: it includes a *spatial frame*, the shifting scene of the action; the *story space*, the composite of all the spatial frames and other locations mentioned in the text; the *story world*, the story space as constructed the reader's knowledge of and experience with the real world; and the *narrative universe*, the world presented as actual by the text plus all of the counterfactual worlds encompassed in the characters beliefs, ideas, fears, or speculations.[17] This spatial world, the scene of the story, is fundamental to what happens in it, as the story's actions are always consistent with its setting.[18]

Narratives, then, are not only inscribed on spatial objects, they are in fact spatially situated, and as readers, we construct mental maps to keep us oriented within the narrative world. It is a natural response because humans organize space by using the body as a reference point, and our language contains numerous words—up, down, left, right, back, forward, etc.—that keep us oriented physically to the environment we are in, whether actual or created in text. Indeed, an increasingly large body of thought known as embodied cognition suggests that what we know cannot be separated from what our bodies perceive as we move through space.[19] But we also orient ourselves in other ways, such as with spatial markers that represent a symbolic geography governed by beliefs, ontologies, or otherwise created relationships—sacred and profane, colonizing and colonized, or town and country, for instance. Narrative arises when a character or event crosses these symbolically charged spaces. Yet another set of spatial markers involves perspective, the point of view, which may refer to characters within the story, the narrator, or the reader.[20]

The point is clear, at least from narrative theory: the stories we construct are inherently spatial. Given this circumstance, what sorts of spatial stories are we telling? Deconstructionists have argued that narrative is so basic to our cultural beliefs that our stories bear little resemblance to reality. "The past is not an untold story," Louis Mink once noted; we are so trapped within our narrative discourse that we have no way to reach a past that exists outside of our cultural assumptions.[21] The German philosopher Martin Heidegger offered another view, one more attractive to humanists who are uncomfortable with the postmodern divorce between stories and an inherent reality. In this view, we have no other way to experience reality other than through our narratives; it is a natural human impulse, as a result, "we live in an endlessly storied world."[22]

And in this storied world, most of the narratives we construct focus not on space, an abstract geometrical concept, but rather on place, the particular expression of geographical space. Humanists embrace the concept of place because of its power in our individual and collective lives. What is authentic about us—our very identity—is inextricably bound up with the places we claim as our own. As novelist Eudora Welty once observed, "place has a more lasting identity than we have, and we unswervingly attach ourselves to identity."[23] Even if we wanted to, we cannot escape place. "The world comes bedecked in places," philosopher Edward Casey has noted. "It is a place-world to begin with."[24]

The concept of place appeals to humanists, who see it as the carrier of culture, which implies that no one class or group controls it. Instead, places are particular and inclusive; they are "organized worlds of meaning," characterized by experience, emotion, and memory.[25] A place exists in past, present, and future time, whereas a space exists only in the present. In fact, place blends spatial and temporal characteristics, a quality we designate as spatiotemporal or, as geographer Doreen Massey has termed it, time-spaces, by which she means the essential conflation of the attributes that define places as distinctive.[26] We confound time and distance when we discuss space—sometimes we refer to time as a surrogate for the nearness or farness of an object—but we can use either attribute to measure space. Our sense of place depends upon the simultaneous connection of both time and space. One attribute alone, either time or space, is not sufficient to define it. Time as a part of place does not need to be precise or even factually accurate, but, unlike space, which exists independent of time, place exists only in time. Time gives specificity to space. The association with the past creates the particularity that space requires to become place, a meaningful location with three distinct attributes—location (fixed coordinates), locale (the material and visual space associated with a location, or its setting), and a sense of place (the historical and emotional characteristics tied to a defined space).[27]

Geographer Yi Fu Tuan argues that "'space' and 'place' require each other for definition."[28] Although undoubtedly true, this cryptic notion can obscure as well as illuminate because the two concepts, if mirror images of each other, also depend on the meaning attached to them by culture. Take, for example, the different ways white explorers and Indians viewed the land and sea, as recounted by travel writer Jonathan Raban as he retraced Captain George Vancouver's explorations along the Pacific coast of Canada in 1792. Vancouver was bewildered by what he considered the natives' aimless tracks across the open, undifferentiated ocean, not realizing that they were maneuvering among friendly and unfriendly spirits.[29] For the Indians, the ocean was place; for Vancouver, it carried no such cultural meaning and simply was space. In this instance, these concepts of space and place were mirror images within each culture and across cultures as well.

What may not be apparent at first glance is the terms we use are themselves highly contested: space and place are everywhere, and their names and definitions have been legion. In modern practice, space and place have become ideological battlegrounds, divorced from the geography and history that embraces them both. The separation first began at the end of the eighteenth century when, with the emergence of modern science, place became transformed into location or, simply, position, a spatial notion. In

Newtonian physics, space is abstract until it can be fixed, a characteristic that allows it to be measured and verified, thus giving it value within scientific method. It can be evaluated with other points for patterns that reveal a universal law. Places offered variations that were not interesting except as local cases; the pattern among them was what was interesting and useful.

Humanists found this meaning of place especially unappealing. During the nineteenth century, scholars began talking about the sense of place, often in romantic terms, as a way to understand the value of the local. By the late-twentieth century, the humanistic revolt against the space-place distinction of the Newtonian model was in full force, with Edward Casey arguing convincingly that the experience of place actually precedes knowledge of space. Resurrecting an ancient insight that it is through the experience of place that we perceive the world, Casey concludes that humans are "ineluctably place-bound." We are, in his playful phrase, "more even than earthlings, we are placelings." Far from being particular, place is "something general, perhaps even universal."[30] Place-making is not an event to be measured; rather, it is a process of being in a "configurative complex of things."[31] We are continually making place by our acts of living in space. Massey has even argued that place is an open and hybrid concept, a product of interconnecting flows and not something that is rooted or fixed.[32] We invest these places with meaning, but they do not exist as isolated and independent spaces. Instead, they are both general and particular. We all experience place, but the places we experience are all different. The world, in brief, is diverse and complex, and we can understand it only through an appreciation of the uniqueness of places and the events and cultures that they hold.

The notion of diversity and complexity tracks well with humanists' view of reality as weblike, to use philosopher Michael Oakeshott's phrase, in which everything is related in some way to everything else. Interdependency is the lingua franca of the humanities, and most recently it has become embodied in practice theory, or in the view of one of its leading proponents, historian William Sewell, "social life may be conceptualized as being composed of countless happenings or encounters in which persons or groups of persons engage in social action." In this view, societies and social systems are "continually shaped and reshaped by the creativity and stubbornness of their human creators."[33] Another historian, Ed Ayers, has labeled this concept "deep contingency," an effort to understand society as a whole with "all structures put into motion and motion put into structures."[34]

The problem comes when we try to use a linear form, the written narrative, to capture this complex reality. It is not simply that we cannot mimic with words the chaos or simultaneity of lived existence—Thomas Carlyle lamented in 1830 that an observation is successive in its recounting, whereas "things done were not in a series but a group"[35]—but that we construct our stories with an end in mind. The narrative carries with it a teleological imperative to explain events as a consequence of past actions or causes and to derive some meaning from the story, lessons that we can use in some way to understand who we are. Our aim, then, is not to reveal fluid and dynamic deep contingencies per se but rather to advance an argument. Writing does this superbly by linking action to consequence, cause to effect. But its linear structure also forces a narrowing or a selectivity among the evidence we use to fashion our account, no matter how thick our descriptions or how evocative our language, thereby giving truth to poet William Blake's lines:

> If the doors of perception were cleansed, everything would
>   appear to man as it is: infinite.
> For man has closed himself up, till he sees all things
>   thro' narrow chinks of his cavern.[36]

Mapping, the tool long favored by geographers, offers a different way to achieve the goal of capturing complexity. It is visual and integrative in a way that words cannot mimic. It also comports well with the aims and methods of humanists. Representation of the past, suggests historian John Lewis Gaddis, is a kind of mapping where the past is a landscape and history is the way we fashion it. The metaphor, one consistent with disciplinary traditions across the humanities, makes the link between "pattern recognition as the primary form of human perception and the fact that all history ... draws upon the recognition of such patterns."[37] In this sense, mapping is not cartographic but conceptual. It permits varying levels of detail, not just as a reflection of scale but also of what is known at the time. Like the map, history becomes better and more accurate as we continue to accumulate more detail, observe its patterns, and refine our knowledge. But mapping, except as metaphor, requires a map, which David Harvey notes, "are typically totalizing, usually two-dimensional, Cartesian, and very undialectical devices."[38]

Does this mean then that we remain in the same position we currently occupy, forced to choose between competing narratives or maps? Or do geospatial and other digital technologies offer a means of avoiding our dilemma through a medium that encourages multiplicity, competing perspectives, and alternate worldviews?[39] Traditional GIS with its emphasis on precise measurement and spatial models is not the solution; neatly categorizing geographic complexity into entities, fields, and objects contrasts starkly with the humanities emphasis on ambiguity, complexity, nuance, and plurality. To this end, GIS cientists have made recent advances in spatial multimedia, GIS-enabled web services, geovisualization, cyber geography, and virtual reality, among other tools, that provide capabilities far exceeding the abilities of GIS on its own. Geo-visualization especially offers an approach useful to humanists because its arena is mental representations, not maps of precisely measured objects; it also aims not to chart what is known but rather to explore what is unknown.[40] Collectively, these technologies also allow us to probe the situated knowledge that resides in dynamic and contested memories and to understand what Stuart Aitken has called the affective or emotional geographies of space and place.[41] They have the potential, in brief, to revolutionize the role of place in the humanities by moving beyond the two-dimensional map to explore dynamic representations and interactive systems that will prompt an experiential, as well as rational, knowledge base.

Humanists work largely with texts, so a key challenge is to learn to frame narratives about individual and collective human experience that are spatially contextualized. Not only is the vast bulk of human experience recorded as text rather than numbers, words are the preferred medium of both ordinary and scholarly communication, regardless of topic or field. Finding ways to make the interaction among words, location, and quantitative data more dynamic and intuitive will yield rich insights into complex sociocultural, political, and economic problems, with enormous potential for areas far outside the traditional orbits of humanities research. All texts exist in both metaphorical and geographic space, so textual analysis itself may be framed by reciprocal transformations from text to map and map to text, as suggested by the novel mappings and *spatial imaginaire* of cultural and literary studies.[42]

We already have hints of what development along these lines can offer to the humanities. Within the field of cultural heritage, archaeologists have used GIS and computer animations to reconstruct the Roman Forum, for example, creating a three-dimensional world that allows users to walk through buildings that no longer exist, except as ruins. We can experience these spaces at various times of the day and seasons of the year. We see more clearly a structure's mass and how it clustered with other forms to mold a dense urban space. In this virtual environment, we gain an immediate, intuitive feel for proximity and power. This constructed memory of a lost space helps us recapture a sense of place that informs and enriches our understanding of ancient Rome.[43] A similar, although more ambitious, project uses laser-scanning technology (LiDAR—light detection and ranging) to create three-dimensional models of major heritage sites and allows scholars and others to roam this virtual environment at will.[44] Elsewhere, scholars are mining texts for spatial references and using them to repopulate the Dublin of James Joyce's *Ulysses* or William Shakespeare's London with the sense of life and possibilities embedded in the past, what Paul Carter has called "intentional history."[45]

Virtual environments are only one way that GIScience is moving us beyond conventional uses of GIS and mere mapping. The geospatial web also holds considerable potential for the spatial humanities to automate the identification and mapping of people, events, places, and spatial relationships from textual resources. Similarly, the ability to transform unstructured text into structured maps suggests that maps may become portals into narratives rather than illustrations of what scholars have written, making spatially aware semantic connections among data and moving toward more complex forms of spatiotextual analysis. Spatialization techniques such as self-organizing maps and text clouds identify clusters in text documents that share similar characteristics in both geographical and metaphorical space. Text-to-map transformations reflect both absolute and relative space by extracting spatial relationships embedded in text and then using this information to go beyond strict map making.[46] Gaming engines also offer ways to reconceptualize the role of space in the humanities by privileging agent-based exploration rather than linear movement as a means of discovery. Spatial stories in this environment are "held together by broadly defined goals and conflicts and are pushed forward by the character's movement," not by the structure of an argument.[47] Truth and authenticity are measured not by standards of causality but by the game's ability to "conjure up experiences of space" that expand and improve our understanding of a complex and multifaceted reality.[48]

These developments from GIScience and related disciplines are important because they quickly are moving us beyond the constraints of a positivist technology that is ill-suited for much humanities research. The same cannot be said with as much assurance about time, which has long been the central lens through which historians view change. The spatial and temporal turns (the New Historicism) go together, and it is unwise to separate the two or prioritize one over the other. GIS has struggled to adequately handle the complexities of these spatiotemporal needs, with the result that the software emphasizes space and treats time as categorical and discontinuous. But the spatial humanities require both time and space: to speak of history as dealing with time, and geography with space, is too simplistic a divide. Massey's idea of exploring multiple trajectories through space and time[49] is much more suited to humanities research, but GIS struggles to provide an environment in which this integrated space-time can be explored profitably.

Central to the emergence of the spatial humanities is a trust that the contingent, unpredictable, and ironic in history and culture can be embodied within a context that incorporates space alongside of time. What we require is a spatial narrative that acknowledges how engaged human agents build spatially framed identities and aspirations out of actions, behaviors, imagination, and memory. At its core, this narrative focuses on spatial patterns as a means of understanding social interaction. It reflects a geography of the constant interaction between structure and process, a continuous interplay between society and the individual and/or group within a spatial environment that both shapes and is shaped by social norms and by individual or group agency. This narrative also must accommodate time and contingency; the social interactions influenced by and influencing space represent, in fact, a web of choices, and the narrative becomes a braided thread (or multiple threads) of those choices over time. But the real question is not the definition of a spatial narrative but how to tap digital and spatial technologies to move narrative beyond the linear constraints of written language into a more fluid and reflexive process in which we can see and experience change and development as a way of understanding an event or a place more fully. If the current scholarly interest in networks is understood as an initial foray into the analysis of important interactions, the spatial narrative can be envisioned as a much richer and complex presentation, one that is geared to the analysis of vast datasets and undertaken in such a way as to maximize experimentation with evidence of contingency, contradiction, and change over time.[50]

Here is where the deep map becomes important. Stemming from the affective stance of unitary urbanism and psychogeography associated with the Situationist International in 1950s France, this approach "attempts to record and represent the grain and patina of place through juxtapositions and interpenetrations of the historical and the contemporary, the political and the poetic, the discursive and the sensual."[51] What is required, the Situationists argued, is an understanding "of the specific effects of the geographical environment, consciously organized or not, on the emotions and behavior of individuals."[52] The idea of deep mapping has a counterpart in geography in the work of Yi Fu Tuan's *Topophilia: A Study of Environmental Perception, Attitudes and Values*,[53] which proposed exploring the connectedness and ties between human emotion and the physical fabric of landscape. As a new creative space, deep maps have several qualities well-suited to a fresh conceptualization of GIS and other spatial technologies as they are applied to the humanities. They are visual, time-based, and structurally open; they are inherently unstable, continually unfolding and changing in response to new data, new perspectives, and new insights. Their aim is not objectivity or authority but rather a negotiated conversation between insiders and outsiders, experts and contributors, over what is represented and how. In their essence, deep maps are the means by which we represent the contested meanings of space and place, as well as the dynamics that produce them. They are curated products—reflexive and self-conscious, historically informed, yet critical and alert to the politics of mapping. As such, they contain the seeds of their own subversion.

The similarity between a deep map and advanced spatial technologies seems evident. Traditional GIS operate as a series of layers, each representing a different theme and tied to a specific location on planet earth.

These layers are transparent, although the user can make any layer or combination of layers opaque while leaving others visible. The advent of light-field photography offers another helpful analogue: it uses hundreds of microlenses to capture all the visual information in a scene, with the photographer bringing into focus later whatever parts of the image that interest her, thus permitting multiple perspectives on an event or object.[54]

A deep map of heritage and culture, centered on memory and place, ideally would work in a similar fashion. The layers of a deep map need not be restricted to a known or discoverable documentary record but could be opened, Wiki-like, to anyone with a memory or artifact to contribute. However structured, these layers would operate as do other layers within a GIS, viewed individually or collectively as a whole or within groups, but all tied to time and space that provide perspectives on the places that interest us. It is an open, visual, and experiential space, immersing users in a virtual world in which uncertainty, ambiguity, and contingency are ever present but all are capable of being braided into a narrative that reveals the ways in which space and time influences and is influenced by social interaction. This space is one in which both horizontal and vertical movement is possible, with the horizontal providing the linear progression we associate with rational argument and vertical movement providing the depth, texture, tension, and resonance of experience.[55] In the deep map, we understand space and place as the product of interrelationships, coexistence, and process, always changing and always in the state of becoming. It permits us to see what literary theorist Raymond Williams has termed a "knowable community," in which what is known is not only a function of objects but also of subjects and observers, of what is desired and needs to be known.[56] It also defines place, in Christopher Tilley's phrase, as "sedimented layers of meaning," the accumulation over time of the events and actions that have happened in a particular space.[57]

Yet deep mapping is not an end in itself. Ultimately, we must use it to construct a spatial narrative and embed within it an argument, which is, in sum, the lingua franca of scholarship. What, then, are the requirements for this narrative? Although the structure may vary widely, it must reveal the influence of real and constructed space on human society and culture, as well as how cultural and social forms shape our understanding of space and place. It must acknowledge that actions occur not simply in time but in space and that events—and our interpretation of them—stem from the confluence of space and time as exposed within a place. This narrative, as a result, must be sensitive to scale: we cannot causally impute local consequences, for instance, to behaviors or events in play across a region or a nation. It also should be alert to the basic ways in which we analyze space—movement, direction, proximity, and connection, among others—and discover the patterns that occur within and among places. And without falling into the trap of spatial determinism, the narrative must reveal how space and place matter to our understanding of society and culture. It must develop a sophisticated argument that gives space-time its own agency, a recognition that we all are place-bound and that the world, in fact, is not flat but endlessly varied and local.

How we construct these narratives will depend, in part, on the richness of our evidence and the tools at our command, but deep mapping can be an ideal storyboard for humanists. It goes beyond traditional uses of GIS and seeks to capture the essence of place and a humanistic sense of distance, direction, and identity. It moves the user from the GIS world of observation to one of experience and enables scholars to engage the material world rather than observe it, as well as to understand more completely how people both create their material world and, in turn, are created by it. Grounded in experiential as well as objective space, deep mapping will provide a representation of society and culture, past and present, with all its rich contradictions and complexities. It will, above all, be a conceptual, technological, and spatial framework that is sensitive to the needs of scholars to tell stories about place.

NOTES

1. A number of recent titles explore this spatial turn in the humanities, including: David J. Bodenhamer, John Corrigan, and Trevor M. Harris, eds., *The Spatial Humanities: GIS and the Future of Humanities Scholarship* (Bloomington: Indiana University Press, 2010); Michael Dear, Jim Ketchum, Sarah Luria, and Douglas Richardson, eds., *GeoHumanities: Art, History, Text at the Edge of Place* (New York: Routledge, 2011); Stephan Daniels, Dydia DeLyser, J. Nicholas Entrikin, and Douglas Richardson, eds., *Envisioning Landscapes, Making Worlds* (New York: Routledge, 2011); Ian N. Gregory and Paul S. Ell, *Historical GIS: Technologies, Methodologies, and Scholarship* (Cambridge: Cambridge University Press, 2008). Also see Peter Doorn, "A Spatial Turn in History," *GIM International* 19, no. 4 (April 2005), http://www.gim-international.com/issues/articles/id453-A_Spatial_Turn_in_History.html (accessed September 12, 2013).

2. For more on this earlier spatial turn, see Stephen Kern, *The Culture of Time and Space, 1880–1918* (Cambridge: Harvard University Press, 1983).

3. Tim Cresswell, *Place: A Short Introduction* (Malden, Mass.: Wiley-Blackwell, 2004) offers a good brief introduction to the postmodern construction of place. Also see Karen Haulttunen, "Groundwork: American Studies in Place," *American Quarterly* 58 (March 2006), 1–15; Richard Biernacki and Jennifer Jordan, "The Place of Space in the Study of the Social," in Patrick Joyce, ed., *The Social in Question: New Bearings in History and the Social Sciences* (New York: Routledge, 2002), 133–150; Peta Mitchell, *Cartographic Strategies of Postmodernism: The Figure of the Map in Contemporary Theory and Fiction* (New York: Routledge, 2008), 23–26.

4. Anthony Giddens, *The Constitution of Society: Outline of the Theory of Structuration* (Oxford: Oxford University Press, 1984), 364.

5. Michel de Certeau reminds us that "space occurs as the effect produced by the operations that orient it, situate it, temporalize it, and make it function as a polyvalent unity of conflictual programs or contractual proximities." And stories are the constructive means we use to transform spaces into places or places into spaces. See Michel de Certeau, *The Practice of Everyday Life* (Berkeley: University of California Press, 1984), 117–118.

6. See Anne Kelly Knowles, ed., *Placing History: How Maps, Spatial Data, and GIS Are Changing Historical Scholarship* (Redlands, Calif.: Esri Press, 2008) for a good sample of the application of GIS to various topics in the humanities. Also see the special issue of *The International Journal of Humanities and Arts Computing* 3, nos. 1–2 (2009), which is devoted to the use of GIS in a number of humanities disciplines.

7. Paul de Man's quote is available at http://chronicle.com/blogPost/how-theory-damaged-the-humanities/6178 (accessed October 27, 2013).

8. Holocaust historian Saul Friedländer captured this aim succinctly: "Commentary," he suggests, "should disrupt the facile linear progression of the narration, introduce alternative interpretations, question any partial conclusion, withstand the need for closure.... Such commentary may introduce splintered or constantly recurring refractions of a traumatic past by using any number of different vantage points." Saul Friedländer, "Trauma, Transference and Working-Through," *History and Memory* 4 (1992), 39–55.

9. Louis Mink, "Narrative Form as a Cognitive Instrument," in Brian Fay, E. O. Golob, and Richard T. Vann, eds., *Historical Understanding* (Ithaca, N.Y.: Cornell University Press, 1987), 22; Frank Ankersmit, "Historiography," in David Herman, Manfred Jahn, and Marie-Laure Ryan, eds., *Routledge Encyclopedia of Narrative Theory* (London: Routledge, 2005), 217–221.

10. Adam Gopnik, "Inquiring Minds: The Spanish Inquisition Revisited," *The Atlantic*, December 2011, 73.

11. Aristotle, *Poetics*, in *The Complete Works of Aristotle: The Revised Oxford Translation*, ed. Jonathan Barnes, 2 vols. (Princeton, N.J.: Princeton University Press, 1984), 2:2321.

12. Edward Branigan, *Narrative Comprehension and Film* (London: Routledge, 1992), 53.

13. Robert Scholes, James Phelan, and Robert Kellogg, *The Nature of Narrative*, rev. and expanded (New York: Oxford University Press, 2006), 3–16.

14. H. Porter Abbott, *The Cambridge Introduction to Narrative* (Cambridge: Cambridge University Press, 2002), 11.

15. M. M. Bakhtin, *The Dialogic Imagination: Four Essays*, ed. Michael Holquist, trans. Caryl Emerson and Michael Holquist (Austin: University of Texas Press, 1981), 278.

16. Susan Stanford Friedman, "Spatialization: A Strategy for Reading Narrative," *Narrative* 1, no. 1 (Jan. 1993), 12–13.

17. Marie-Laure Ryan, "Space," in Peter Hühn et al., eds., *The Living Handbook of Narratology* (Hamburg: Hamburg University), http://www.lhn.uni-hamburg.de/ (accessed May 7, 2014).

18. Kenneth Burke, *A Grammar of Motives* (Berkeley: University of California Press, 1969), 6–7; William Cronon, "A Place for Stories: Nature, History, and Narrative," *Journal of American History* 78, no. 4 (March 1992), 1347–1376.

19. Barbara Tversky, "Spatial Cognition: Embodied and Situated," in Philip Robbins and Murat Aydede, eds., *Cambridge Handbook of Situated Cognition* (Cambridge: Cambridge University Press, 2009), 201–217; Barbara Tversky and Bridgette M. Hard, "Embodied and Disembodied Cognition: Spatial Perspective-taking," *Cognition* 110, no. 1 (2009), 124–129.

20. Sabine Buchholz and Manfred Jahn, "Space in Narrative," in *Encyclopedia of Narrative Theory*, 550–555.

21. Mink, "Narrative Form as a Cognitive Instrument," 22.

22. William Cronon, "A Place for Stories," 1368; Stuart Elden, *Mapping the Present: Heidegger, Foucault, and the Project of Spatial History* (London: Continuum, 2001), 93–111.

23. Eudora Welty, "Place in Fiction," *The Eye of the Storm: Selected Essays and Reviews* (New York, 1990), 119. Roberto Maria Dainotto discusses the postcolonial connection of place and literature in "'All the Regions Do Smilingly Revolt': The Literature of Place and Region," *Critical Inquiry* 22, no. 3 (Spring 1996), 486–505.

24. Edward S. Casey, "How To Get from Space to Place in a Fairly Short Stretch of Time: Phenomenological Prolegomena," in S. Feld and K. H. Basso, eds., *Senses of Place* (Santa Fe, N.M.: School of American Research Press, 1996), 43.

25. Philip J. Ethington, "Placing the Past: 'Groundwork' for a Spatial Theory of History," *Rethinking History: The Journal of Theory and Practice* 11, no. 4 (Dec. 2007), 481. The quote is from Yi Fu Tuan, *Space and Place: The Perspective of Experience* (Minneapolis: University of Minnesota Press, 1977), 179.

26. Doreen Massey, *For Space* (London: Sage, 2005), 177–180.

27. Tim Cresswell, *Place: A Short Introduction* (London: Blackwell Publishers, 2004), 7–8.

28. Tuan, *Space and Place*, 6.

29. Jonathan Raban, *Passage to Juneau: A Sea and Its Meanings* (New York: Pantheon Books, 1999), 103, as cited in Cresswell, *Place*, 11.

30. Casey, "How To Get from Space to Place in a Fairly Short Stretch of Time," 19.

31. Ibid., 25.

32. Doreen Massey, "A Global Sense of Place," in Trevor Barnes and Derek Gregory, eds., *Reading Human Geography* (London: Hodder Arnold, 1997), 315–323.

33. William H. Sewell, Jr., *Logics of History: Social Theory and Social Transformation* (Chicago: University of Chicago Press, 2005), 110–111.

34. Edward L. Ayers, "Turning toward Space, Place, and Time," in Bodenhamer, et al., *The Spatial Humanities*, 7.

35. Thomas Carlyle, "On History," in Fritz Stern, ed., *The Varieties of History: From Voltaire to the Present* (New York: Vintage Books, 1972), 95.

36. William Blake, "The Marriage of Heaven and Hell," in David V. Erdman, *The Complete Poetry and Prose of William Blake*, electronic ed. (Institute for Advanced Technology

in the Humanities, University of Virginia. 2001), http://www.blakearchive.org/blake/erdman.html (last accessed October 27, 2013).

37. John Lewis Gaddis, *The Landscape of History: How Historians Map the Past* (New York: Oxford University Press, 2002), 33.

38. David Harvey, *Justice, Nature and the Geography of Difference* (Oxford: Blackwell, 1996), 4.

39. See Claudio Fogu, "Digitalizing Historical Consciousness," *History and Theory* 47 (May 2009), 103–121.

40. Alan M. MacEachren, Mark Gahegan, and William Pike, "Visualization for Constructing and Sharing Geo-scientific Concepts," *Proceedings of the National Academy of Sciences* 101, no. 1 (2004), 5279–5286.

41. James Craine and Stuart Aitken, "The Emotional Life of Maps and Other Visual Geographies," in Martin Dodge, Rob Kitchin, and Chris Perkins, eds., *Rethinking Maps: New Frontiers in Cartographic Theory* (New York: Routledge, 2009), 168–185.

42. Franco Moretti, Graphs, Maps, Trees: Abstract Models for a Literary History (London: Verso, 2005); Barbara Piatti, "Mapping Literature: Toward a Geography of Fiction," ftp://cartography.ch/pub/pub_pdf/2009_Piatti_Geography_of_Fiction.pdf (last accessed May 7, 2014).

43. Digital Roman Forum Project, http://dlib.etc.ucla.edu/projects/Forum/ (last accessed October 27, 2013).

44. Information about this use of LiDAR is available at http://www.ted.com/talks/ben_kacyra_ancient_wonders_captured_in_3d.html (last accessed October 27, 2013).

45. Paul Carter, *The Road to Botany Bay* (London: Faber and Faber, 1987), 3.

46. Trevor M. Harris, L. Jesse Rouse, and Susan Bergeron, "The Geospatial Semantic Web, Pareto GIS, and the Humanities," in Bodenhamer, et al., *The Spatial Humanities*, 124–142; May Yuan, "Mapping Text," in Bodenhamer, et al., *The Spatial Humanities*, 109–123; Ian N. Gregory and Andrew Hardie, "Visual GISting: Bringing Together Corpus Linguistics and Geographical Information Systems," *Literary and Linguistic Computing* 26 (2011), 297–314. Also, see Edward L. Ayers, "The Pasts and Future of Digital History," http://www.vcdh.virginia.edu/PastsFutures.html (last accessed October 27, 2013).

47. Henry Jenkins, "Game Design as Narrative Architecture," http://web.mit.edu/cms/People/henry3/games&narrative.html (last accessed October 27, 2013). Also see Connie Veugen and Felipe Quérette, "Thinking out of the Box(and back in the Plane): Concepts of Space and Spatial Representation in Two Classic Adventure Games," *Eludamos. Journal for Computer Game Culture* 2, no. 2 (2008), http://www.eludamos.org/index.php/eludamos/article/view/vol2no2-6/86 (last accessed October 27, 2013).

48. Christoph Classen and Wulf Kansteiner, "Truth and Authenticity in Contemporary Historical Culture," *History and Theory (Theme Issue)* 47 (2009), 3.

49. Doreen Massey, *For Space*, 9–15; Doreen Massey, "Space-Time, 'Science' and the Relationship between Physical Geography and Human Geography," *Transactions of the British Geographical Society, New Series* 24 (1999), 261–276. A useful discussion also may be found in Donna Peuquet, *Representations of Space and Time* (New York: The Guildford Press, 2002).

50. See John Corrigan, Trevor M. Harris, and David J. Bodenhamer, "The Spatial Humanities" (white paper, National Science Foundation, Directorate for Social, Behavioral and Economic Sciences, 2010 [SBE 2020: ID 163]), http://www.nsf.gov/sbe/sbe_2020/submission_detail.cfm?upld_id=163 (last accessed October 27, 2013).

51. Mike Pearson and Michael Shanks, *Theatre/archaeology* (London: Routledge, 2001), 64–65. Also see David J. Bodenhamer, "Creating a Landscape of Memory," *International Journal of Humanities and Arts Computing* 1, no. 2 (2008), 97–110.

52. "Definitions," in Ken Knabb, ed., *Situationist International Anthology* (Berkeley: Bureau of Public Secrets, 1981), 45, as cited in Mitchell, *Cartographic Strategies of Postmodernity*, 119n44.

53. Yi Fu Tuan, *Topophilia: A Study of Environmental Perception, Attitudes and Values* (1974; repr., New York: Columbia University Press, 1990).

54. Rob Walker, "Deep Focus: How a New Camera Will Revolutionize Photography," *The Atlantic*, December 2011, 16. Although it sounds futuristic, such a camera, the Lytro, appeared in 2012. Also see Peta Mitchell, "'The Stratified Record upon Which We Set Our Feet': The Spatial Turn and the Multilayering of History, Geography, and Geology," in Dear, et al., *GeoHumanities*, 71–83.

55. For another view of how the use of Web 2.0 tools will reshape traditional notions of authority, see Ian Johnson, "Spatiality and the Social Web: Resituating Authoritative Content," in Dear, et al., *GeoHumanities*, 267–276. Also see the various essays in Arno Scharl and Klaus Tochtermann, *The Geospatial Web: How Geobrowsers, Social Software, and Web 2.0 Are Shaping the Network Society* (London: Springer, 2007).

56. Raymond Williams, *The Country and the City* (New York: Oxford University Press, 1973), 165.

57. Christopher Tilley, *A Phenomenology of Landscape: Places, Paths and Monuments* (Oxford: Berg Publishers, 1994), 27.

2

DEEP GEOGRAPHY—DEEP MAPPING

Spatial Storytelling and a Sense of Place

TREVOR M. HARRIS

Maps are more than pieces of paper. They are stories, conversations, lives and songs lived out in a place and are inseparable from the political and cultural contexts in which they are used.

INTRODUCTION

Deep mapping, spatial storytelling, and spatial narratives are incomplete terms struggling to capture and imbue meaning to abstract thoughts of a more profound, insightful, reflexive, multimedic, perhaps quixotic representation of humanistic space than currently prevails. To focus on the term *deep mapping* emphasizes William Least Heat-Moon's[1] and Barry Lopez's[2] admonition to throw away the "wrong sort" of superficial "thin" maps and to seek a "less suspicious" deep map, but it also encourages an emblematic continuation of a mapped representation of space. Spatial storytelling and spatial narratives provide rich connotations and attachments to the humanities' storied-narrative style, though, despite a potential gamut of methodological approaches, narratological representations of space continue to seek a conceptual core. For a number of reasons then, I opt for the term *deep geography* to encompass deep mapping, spatial storytelling, and spatial narrative and to situate these themes within a conceptual and methodological framework that can be brought to bear to define, refine, and explore these nebulous concepts.

To counteract the magnetic allure of addressing the "how to" element of implementing manifestations of these terms, themes, and ideas, it is important to first understand the critical question as to "why" the concepts, techniques, practices, and usages of deep geography are important and the purpose(s) in pursuing them. I suggest that the ultimate goal is to explore and attain a deeper understanding of place, as distinguishable from that of space. Place and sense of place, place-making, and experiencing place are well-established fields within geography and deep mapping links these to humanistic examinations of deep contingency. Current approaches to examining place and especially historical place, draw upon advances in local history, microhistory, historical geography, cartography, and geographic information systems (GIS). GIS offers considerable potential for the spatial humanities but the challenge is to shift from a view of humans as entities or data points to an examination of behavior, the material and imaginary worlds, and the relationships that compose notions of a nuanced, nonreductionist, deeply contingent, and scaled conception(s) of place.[3]

I suggest a number of approaches and possible methods to examine deep geography that draw from a smorgasbord of qualitative GIS, ethnography, virtual environments, storyboarding, critical GIS and participatory GIS (PGIS), and neogeography and its related component parts of the geospatial web, crowd sourcing, and volunteered geographic information (VGI). I also suggest that several metaphors can provide additional insight and motivation as to possible methodological implementations of deep mapping and spatial storytelling.

THIN MAPS

Use of the term *deep map* requires unpacking as it relates to some affective implied association between the cartographic map and the mapping process and contingent humanistic behavior. Maps have come to represent the embodiment and outward visual expression of the geography of an area. While not always limited to the cartographic mapping of geographical space, the act of mapping features to convey information is embedded in our everyday consciousness and usage. Maps have intrinsic characteristics and properties that have contributed to their success as a communication medium, for they possess a considerable capacity to store and display vast quantities of spatial data on a single sheet or image. Furthermore, the graphical expression of often complex spatial information can be rapidly and intuitively interpreted because the human eye is a very effective and powerful image processor.

Traditional map content tends to be dominated by the physical fabric of society and, as bureaucratic and management tools, has been used extensively to map infrastructural features such as roads, railways, and transmission lines; physical features such as the terrain, hydrography, and vegetation cover; places of population and settlement; and a host of other features traditionally found on a topographic map sheet. These cartographic products are largely intended for general audiences. Thematic maps draw more heavily on choropleth mapping and other cartographic representations and focus on specific themes that often include cultural and social data. Although challenged by the plethora of mapping opportunities now available to non–cartographically trained map creators enabled by the broader availability of GIS and the geospatial web, these cartographic representations are steeped in an extensive praxis of design, aesthetics, science, and technology that ostensibly seek to model aspects of reality and communicate that information to a, usually defined, audience.[4] While Robinson[5] would suggest that good map design is premised on having an intended audience in mind, it is MacEachren[6] who stresses that a good map should imply authenticity and show multiple elements that reinforce relationships in the map and thus its value and significance. These cartographic products are characterized by a focus and concern for precision and accuracy. These maps can be construed as thin maps in that they are conceived, designed, created, and maintained by experts for both general and specific audiences, often to meet specific governmental or corporate needs, and are heavily focused on the material and physical characteristics of landscape and society. For several reasons, I favor the term *thin map* to describe these cartographic products rather than "shallow map," which is the antonym to deep map. The term *shallow*, when applied to these maps, intentionally or otherwise, implies a meaning of superficiality and inconsequentiality and that they are lightweight and lacking in substance. There is overwhelming evidence to disprove these latter descriptions and any consideration of the impact of GIS on contemporary society validates the value and contribution of these maps across a range of societal endeavors. These maps and their map content have formed the backbone of GIS and national spatial data infrastructures and have proven invaluable in numerous areas of government, business, environment, and community applications.

DEEP MAPS

In contrast to these surficial thin maps, deep maps represent an alternative form of mapping and of content. As Lopez commented:

> I would like to tell you how to get there so that you may see all this for yourself. But first a warning: you may already have come across a set of detailed instructions, a map with every bush and stone clearly marked, the meandering courses of dry rivers and other geographical features noted, with dotted lines put down to represent the very faintest of trails. Perhaps there were also warnings printed in tiny red letters along the margin, about the lack of water, the strength of the wind and the swiftness of the rattlesnakes. Your confidence in these finely etched maps is understandable, for at first glance they may seem excellent, the best a man is capable of; but your confidence is misplaced. Throw them out. They are the wrong sort of map. They are too thin. They are not the sort of map that can be followed by a man who knows what he is doing. The coyote, even the crow, would regard them with suspicion.[7]

The intellectual roots of deep maps lies in a combination of eighteenth-century antiquarian approaches to geography, history, people, culture, and place; in the detailed local histories of historical geographers such as W. G. Hoskins[8]; in the thick description of William Least Heat-Moon's *PrairyErth*[9]; in the insightful observations into deep mapping and performance drama of Pearson and Shank[10]; in Certeau's[11] spatial stories and practices of everyday life; in Yi Fu Tuan's *Topohilia*[12]; and in the early work of the Situationists International[13] and psychogeography.

The short-lived Marxist avant-garde Situationist International of mid-twentieth-century Europe sought to resist advanced capitalism and the perceived spectacle of the fake reality that they claimed masked capitalism's degradation of people's lives by seeking alternatives through the "constructedness of situations."[14] Guy Debord, the principal leader and theorist of Situationist International used the term *situation* to suggest a critical evaluation of everyday life, moments of life, which he suggested are transformed into a superior and self-desired quality. In similar vein to Marx's commodification of the material and the symbolic world, the Situationists asserted that advanced capitalism also commodified experience and perception and that a mass media–induced image-laden reality produced fake models that were at odds with everyday experience. Alternative pathways to creating life experiences were proposed by the Situationists such as psychogeography that sought to evaluate the effects of the geographical environment on the emotions and behavior of people. As Debord wrote in his *Introduction to a Critique of Urban Geography:*

> Geography ... deals with the determinant action of general natural forces, such as soil composition or climatic conditions, on the economic structures of a society, and thus on the corresponding conception that such a society can have of the world. Psychogeography could set for itself the study of the precise laws and specific effects of the geographical environment, whether consciously organized or not, on the emotions and behavior of individuals. The charmingly vague adjective psychogeographical can be applied to the findings arrived at by this type of investigation, to their influence on human feelings, and more generally to any situation or conduct that seems to reflect the same spirit of discovery.[15]

In this regard, psychogeography combines both objective and subjective knowledge and through the theory and action of *dérive* or drift, a technique of rapid passage through varied ambiences, people would let go of conscious actions, relations, and movement and be drawn to the attractions of the terrain and the encounters to be found there.[16] Thus in *unitary urbanism,* the Situationists rejected Euclidian projections of space and suggested that as travelers traversed an urban environment, they would allow themselves to be guided by their emotional responses to the environment and to the encounters within it. Thus the Situationists and the concepts surrounding psychogeography can be seen as both critique and challenge to the current stream of geospatial technology and data production. If, through the extensive mapping of space premised on authenticated spatial data infrastructure and GIS,[17] and the use of mobile location-based services and geotagged VGI, all items in space could be mapped, would it still be possible to understand and experience place? And what of the chance encounters that the ambience of place and placemaking can engender through sound, emotion, smell, contact, interaction, and movement?

Given the privileged role given to thin maps and the mapped world of the physical environment is it possible to know a location even if every object and measured space were mapped? The social theoretic critique of the GIS and Society debates in the 1990s, and more lately critical GIS, posed these very same questions about the emphasis and privileged position given to the Boolean logic of the Turing machine and of positivist GIS and the cartographic metaphor.[18] While we may know the space, how well do we understand the place and movement between places? Does our (thin) mapping of the environment become the essence of the virtual tourist, experiencing place vicariously and by proxy but without the sounds, sights, interactions, encounters, and emotions associated with place?

From the concepts of the Situationist International, the divergence between thin maps and deep maps—between space and place—becomes clearer and more focused. As Debord reflects, "We live in a spectacular society, that is, our whole life is surrounded by an immense accumulation of spectacles. Things that were once directly lived are now lived by proxy. Once an experience is taken out of the real world it becomes a commodity. As a commodity the spectacular is developed to the detriment of the real. It becomes a substitute for experience." The trends being experienced in GIS toward an exhaustive mapping of the physical environment, "the world on your desktop," predominantly reinforce the mapping of the material to the exclusion of the intangible, of emotion and experience, and the contingent human encounter with the physical, symbolic, and imaginary landscapes. The "virtual tourist" takes on particular connotations when the physical is privileged over the human experience, which is transformed to that of data points, entities, and objects. Deep maps reflect the complex interaction of the physical and human environments and their relations and behaviors that are nuanced, nonlinear, branching, and so very difficult to map. In many ways, deep maps seek to "map" the unmappable and therein lies the challenge.

By way of illustration, I liken these space-place and thin map–deep map issues to the role of place names in mapping. The etymology of Native American place names, for example, provides a rich understanding and descriptive use of terms that reflect the physical and cultural worlds that Native Americans inhabited. Certainly physical features such as lakes and mountains are represented, but these are interwoven into morphologically complex and semantically rich forms and tied to stories (both remembered and forgotten) in folklore and the ethnocentric descriptions of places.[19] The polysynthetic character of many Native American languages often enables these complex and semantic nuances to be captured in one lexical term. As Bright suggests, "Of the linguistic artifacts and distinct semantic domains in lexicons of all the world's languages, place names tell us something not only about the structure and content of the physical environment itself but also how people perceived, conceptualized, classified, and utilized that environment."[20] To use a well-known Welsh place name example: *Llanfairpwllgwyngyllgogerychwyrndrobwllllantysiliogogogoch,* which translated means "[St.] Mary's Church (*Llanfair*) [in] the hollow (*pwll*) of the white hazel (*gwyngyll*) near (*goger*) the rapid whirlpool (*y chwyrndrobwll*) [and] the church of [St.] Tysilio (*llantysilio*) with a red cave (*[a]g ogo goch*)." At fifty-eight letters long, this place name is one of the longest in the world and while it serves the purpose of my point here it should be acknowledged that it was an artificially contrived nineteenth-century publicity stunt to attract tourists to the area. Nonetheless, toponyms provide valuable insight into the ways in which humans experience the world and appropriate images of the landscape to describe, interpret, and communicate their experiences of the physical and social environment.[21] As Harrington, who examined the ethnogeography of the Native American tribe of the Tewa, powerfully suggests: "The Tewa have a marked fondness for geographical conversation."[22]

I metaphorically liken these humanized usages of place names to the toponymic equivalent of the deep map and certainly so when made in comparison with other forms of place-naming nomenclature. A comparison to the naming conventions used in Antarctica, for example, takes one to the U.S. Board on Geographic Names.[23] In the largely unexplored and uninhabited land of Antarctica where human encounter with the landscape has largely been a result of exploration and research centers, place names mostly demarcate features or locations based on the names of these very same explorers and "deserving people." Indeed the three-tiered naming conventions are illustrative of this. First-order place names that are allocated to physical features are based on a select group of expedition leaders, discoverers of outstanding significance, contributors to science, and financial backers of expeditions. In the second naming tier come ships' captains, those who exhibited acts of heroism, heads of learned societies, and those who contributed to the acquisition of equipment. Finally, in the third order of name determination come members of expeditions and teachers and administrators who contributed to the training of polar explorers. Interestingly, natural features are similarly tiered with first-order coasts, seas, plateaus, mountain ranges, and glaciers; second-order prominent mountains, capes, gulfs, harbors, straits; and third-order features comprising cliffs, nunataks, minor shore features, anchorages, bays, and coves. The white space of the Antarctic to be found (literally) on the maps of Google Earth or Google Maps, and most other maps and atlases for that matter, are an extreme example of the thin map. The extreme nature of the landscape, which contains the barest elements of geography, settlement, or history, albeit a fascinating geopolitical record, could indeed assume the label of white map, or light map, or thin map.

A second intellectual origin for deep maps lies in eighteenth-century antiquarian approaches to geography and history. Pearson and Shank,[24] for example, suggest that historically, place included history, folklore, natural history, and hearsay, and as such their deep map "attempts to record and represent the grain and patina of a location [through the] juxtapositions and interpenetrations of the historical and the contemporary, the political and the poetic, the factual and the fictional, the discursive and the sensual; the conflation of oral testimony, anthology, memoir, biography, natural history and everything you might ever want to say about a place." In addition, Pearson and Shank provide other conceptual themes that give insight into the character and attributes embedded in deep maps: the *lifeworld,* which is the totality of a person's direct involvement with the places and environments experienced in everyday life; *thick description,* which is the detailed and contextual description of cultural phenomena necessary to discern the complexities behind the action including the codes at work and the possible structures of meaning; *blurred genre,* which is a mixture of scientific practices and narration and an integrated interdisciplinary, intertextual, and creative approach to recording, writing, and illustrating the material past; and finally *scene-of-crime,* which is a cordoned area where anything might be meaningful and potentially constitute evidence.

In providing a methodological framework capable of exploring encounters with landscapes through personal narrative, biography, folklore, text, physical geography, family history, and performance drama, Pearson and

Shank[25] highlight what they consider to be the "affective ties between people and space ... and their cultural and physical environment." The conjoining of encounter and landscape and the use of mixed methods to examine the physicality of site and context and related human emotion and reflection are indicative of the Situationist concepts of Debord. In his text *In Comes I,* Pearson explores the landscape through a first-person narrative as a matrix of related stories where there is no last word on the topic. Using performance as an agent for engaging place, Pearson seeks to make sense of the "multiplicity of meanings that resonate from landscapes and memories; providing a mechanism for enacting the intimate connection between personal biography, social identities and the biography of place—at a variety of scales of rhetoric, within different scales of landscape."[26] Surely, herein lays a powerful analogy and perspective on deep maps.

While the conceptual propinquity of deep maps to performance art and archaeology may at first glance seem ambiguous, Harvie[27] suggests that the combination of geography, natural history, and accounts of the history and lived experience of the inhabitants of a given area are the very basis of deep maps. These aspects are perhaps best seen in the performance work of Brith Gof and his exploration of Welsh identity. Gof insisted on locating performance art in the real and lived environments where the memories that produced those identities first originated. Thus the association of the physical environment with human experience is made explicit through holding performances in "environments of social, industrial, and economic activity (or forced inactivity) which so significantly constituted many people's—especially men's—experience of Welsh identity" rather than in purpose-built theaters.[28] What is clear about thin maps is that while the map content may be rich in specific ways, the inability to disentangle multiple realities represents a serious challenge to its effectiveness as a deep map on its own. Mapping and tracking these multiple pathways, truths, realities, and the multivocality of the deep map is no easy matter to accomplish.

The meta-media Three Landscapes Project with its focus on integrating multiple media including painting, photography, poetry, and performance within cultural and historical geography, archaeology, and art history is a further example of a deep map that assists in teasing out notions of place and identity.[29] In the Three Landscapes Project, the investigators pursue a critique of landscape as a reflection of the relationships among environment, land, place, and history: Three Landscapes. In producing a large 8 feet by 42 feet collage that combines "a variety of mappings, aerial surveys, photographs, journal and journey, with a single figure in the landscape and several orders of text," A Map on a Wall (figures 2.1 and 2.2) gives a graphical emphasis to the creation of a deep map not least through the juxtaposition and layering of materials that represent the complexities of place. As McLucas suggests, the graphic is the "beginning of an attempt to develop new techniques for representing places, peoples, and event—techniques that are both more complex and (dis)located than those associated with the landscape painting, the photograph, or the conventional map."[30]

The hybridity of methods involved in assessing the experiential and the encounter between the tangible physical environment and the intangible symbolic, contested, and imaginary worlds has clear connotations for deep maps. The insightful perspectives on deep maps of the artist Cliff McLucas are significant:

FIGURE 2.1. A Map on a Wall (http://metamedia.stanford.edu/~mshanks/galleries/map-on-a-wall/toc.html).

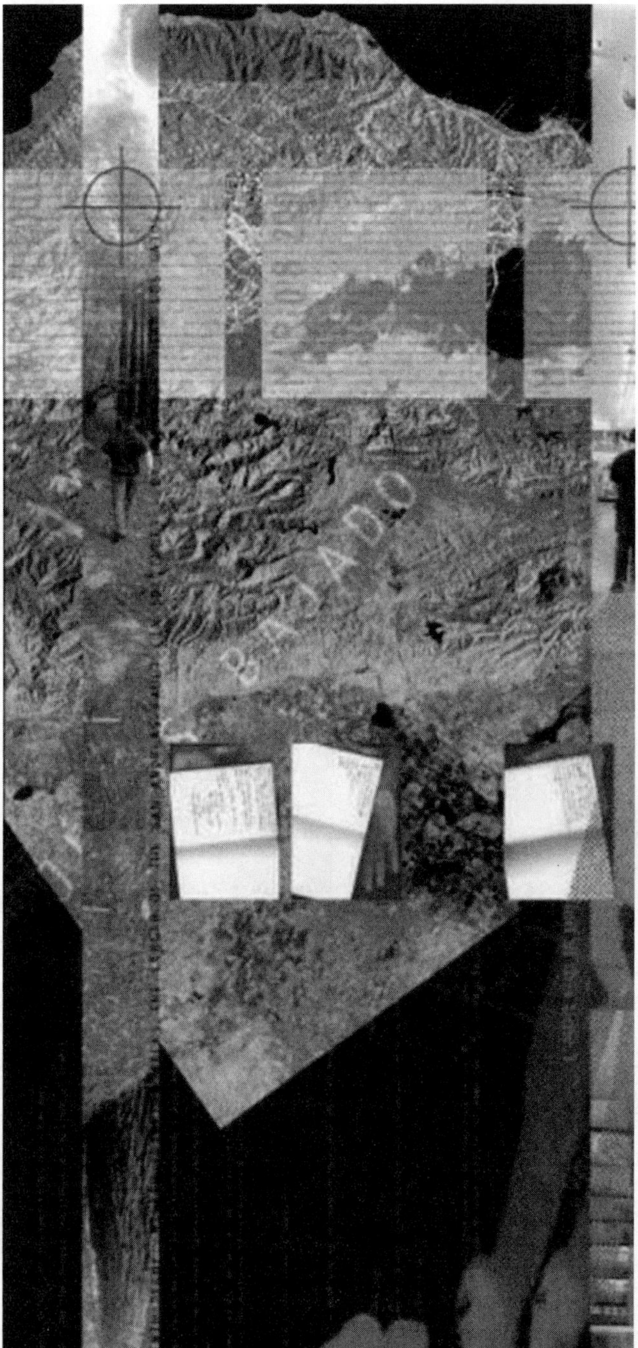
BAJADO

There are ten things that I can say about these deep maps. First, deep maps will be big—the issue of resolution and detail is addressed by size. Second, deep maps will be slow—they will naturally move at a speed of landform or weather. Third, deep maps will be sumptuous—they will embrace a range of different media or registers in a sophisticated and multilayered orchestration. Fourth, deep maps will only be achieved by the articulation of a variety of media—they will be genuinely multimedia, not as an aesthetic gesture or affectation, but as a practical necessity. Fifth, deep maps will have at least three basic elements—a graphic work (large, horizontal or vertical), a time-based media component (film, video, performance), and a database or archival system that remains open and unfinished. Sixth, deep maps will require the engagement of both the insider and outsider. Seventh, deep maps will bring together the amateur and the professional, the artist and the scientist, the official and the unofficial, the national and the local. Eighth, deep maps might only be possible and perhaps imaginable now—the digital processes at the heart of most modern media practices are allowing, for the first time, the easy combination of different orders of material—a new creative space. Ninth, deep maps will not seek the authority and objectivity of conventional cartography. They will be politicized, passionate, and partisan. They will involve negotiation and contestation over who and what is represented and how. They will give rise to debate about the documentation and portrayal of people and places. Tenth, deep maps will be unstable, fragile and temporary. They will be a conversation and not a statement.[31]

The ghost maps of Ethington[32] provide a similar rich and highly detailed visual layering of graphics, representations, texts, and maps that attempt to capture the multiple associations of meanings and inscriptions related to a location and to efforts to recall the past of a place.

SPATIAL STORIES, SPATIAL NARRATIVE, AND GEONARRATIVE

A deep map then is more than a topographical product in that it interweaves physical geography and scientific analysis with biography, folklore, narrative, text, memories, emotions, stories, oral histories, and so much more to contribute to a richer, deeper mapping of space and place. Spatial stories weave pathways through deep maps to track, organize, and record

FIGURE 2.2. A Map on a Wall—an experiment in deep mapping (http://metamedia.stanford.edu/~mshanks/threelandscapes/largemap.html).

people's experiences and relationships with places. The deep map and spatial storytelling take the process of deep mapping further than would an abstract mashup of cultural and physical geographies. If, as Spielberg would suggest, "people have forgotten how to tell a story," how more complex then is it to grapple with the telling of spatial stories? Mark Twain[33] in his harsh criticism of Fenimore Cooper's writings made several suggestions or rules for storytelling, prime among them being:

1. A tale shall accomplish something and arrive somewhere.
2. The episodes of a tale shall be necessary parts of the tale, and shall help develop it.
3. The personages in a tale, both dead and alive, shall exhibit a sufficient excuse for being there.
4. When the personages of a tale deal in conversation, the talk shall sound like human talk, and have a discoverable meaning, and discoverable purpose.
5. Events shall be believable; the personages of a tale shall confine themselves to possibilities and let miracles alone; or, if they venture a miracle, the author must plausibly set it forth so as to make it look possible and reasonable.
6. Finally, the author shall make the reader feel a deep interest in the personages of his tale and their fate; and that he shall make the reader love the good people in the tale and hate the bad ones.

Drawing on these storytelling points, there is a difference to be discerned between deep maps and spatial stories and, I would suggest, between spatial stories and spatial narrative. Some would suggest that storytelling represents the oldest form of human art and was a critical tool in human development and the march toward human understanding. Indeed, when Einstein was asked how best to develop intelligence in young people his response was to say, "Read fairytales. Then read more fairy tales." Storytelling is often simplistically stated as a story having a beginning, middle, and an end; and yet, through these structural elements, there runs a strand or thread that connects the human and the spatial elements of the tale. Film director James Cameron declared that the mind hates fragmentation, so overcoming the chaos of data overload through the provision of some organizing theme such as a spatial story as one traverses a series of events and places is valuable. As Martin[34] suggests, for a good story to be told, the plot or storyline must be clearly present. Stories provide a link between the storyteller and the listener and give structure to events that are channeled toward an end that provides a message and a meaning when the destination is reached. In the telling of a story, the intent is to achieve a specific purpose or an ending which the storyteller has deemed worth recalling. A geographic organizing theme might be a major historical feature such as the national road (U.S. 40) that connects multiple places and stories.[35] The spatial story thus takes a complex flow of events and places and organizes these into an understandable and meaningful form that can be communicated to others. It also elevates the importance of the message and the recounting of events by giving some segments of the deep map greater emphasis, meaning, and purpose. Stories are invariably embedded within a temporal framework, and historians have been adept at using the storytelling and biographical style to appreciate and understand the basis and outcomes of human behavior. Other media outlets including film, theater, art, literature, and music can also use a storied theme to communicate powerful messages.

Spatial stories and spatial narratives might at first blush be considered synonyms, but I suggest that differences exist between a story and a narrative for a spatial narrative is more a neutral recounting of a series of sequential events; a "literal" telling of events. In contrast to a spatial narrative, there are elements to stories that involve the selective arrangement of events and information. As anyone who has read good fiction will know, the plot is a necessary part, but there has to be more than plot to engage the reader and to draw them in. Perhaps the major distinction between spatial story and spatial narrative is that the spatial storyteller will have a point of view which is relayed through story content, writing style, or emotion. The positionality of the storyteller is thus critical in terms of whose story is being told, to whom, and for what purpose, for ultimately the storyteller seeks to deliver a message to the reader. The ability to recount multiple realities and to give voice to multiple groups is a critical element in deep mapping and spatial storytelling for it begins to break down the considerable emphasis given to meta-narratives and to universal truths.

There are obvious similarities with the term *geonarratives* as used by Kwan and Ding.[36] The integrated analysis of narrative text, stories told about peoples' lived experiences of past events which are sequenced within a mixed-methods approach and set within a spatiotemporal setting, helps us understand how GIS might include oral histories, life histories, biographies, and emotions. The style and structure of serious game engines can also contribute significantly to this discussion on spatial stories for they too invariably require the user to create their own self-scripted spatial journey through a usually imaginary world. As with Twain's first rule of storytelling, there is a structured theme running through these "games" whereby the story often unfolds in the form of a quest or journey to achieve a series of goals within the defined geography and resources of the virtual scene. The story is bounded by the geography and by a series of rules that apply to the participants representing a bounded complexity. This structured storytelling stands in contrast to a highly unstructured gaming environment where there are no structures or rules, which produces, in the terminology of serious gaming, a "sandbox" in which users can move freely anywhere and for any purpose.

Spatial stories stem from the universal cultural need to describe, recount, and narrate a particular stream of thought that is situated within, or impacted by, a place or series of places. As with deep maps, a spatial story is enriched by multiple forms of representation and cuts across maps and scales. The narrative voice refers not to layers of information or to media but to the themes, interaction, and experiences that run through such geographies that link the physical and cultural worlds with the fictional, symbolic, and the imaginary. Who determines the content, the strand of spatial thoughts that unites the story and who then owns that story or the information and private things that may have been revealed, represent critical questions in a geocoded world.[37] Given the adversarial and unequal power relations of society, the personal experiences revealed in media-rich deep maps may provide greater diversity of opinion and insight, but in the process, they also become more susceptible to surveillant scrutiny.

Michael de Certeau in his *Spatial Stories: The Practice of Everyday Life*[38] suggests that narration and space cross and connect sites through itineraries: what might be called a spatial trajectory where the stories about places are structured in a linear and interlaced series. Space connects places, but place is also imbued with temporal qualities that spatial stories traverse and organize and link into the story form. I would suggest that John Gay's[39] poem *Trivia: or, the Art of Walking the Streets of London*, a topographical poem describing the perils of walking the streets of London in the eighteenth century, provides a fascinating instance of Certeau's traversing of events and places. Gay's poem refers to the characteristics of the city footmen and ballad singers of London, England, and touches upon such things as the everyday problems of falling masonry, chamber pots being emptied out of windows, overflowing gutters, pickpockets and wig thieves, and what boots to wear:

> Through Winter Streets to steer your Course aright,
> How to walk clean by Day, and safe by Night,
> How jostling Crouds, with Prudence to decline,
> When to assert the Wall, and when resign,
> I sing: Thou, Trivia, Goddess, aid my Song,
> Thro' spacious Streets conduct thy Bard along;
> By thee transported, I securely stray
> Where winding Alleys lead the doubtful Way,
> The silent Court, and op'ning Square explore,
> And long perplexing Lanes untrod before.[40]

In *Trivia*, the reader is taken on a walk through the London of John Gay and reads about contemporary eighteenth-century perspectives on the dirty, crowded, and fascinating street scenes that Gay describes. Using the insights provided by the poem, a twentieth-century historian, geographer, a literary scholar, and gender and costume scholars are collected in the work by Brant and Whyman[41] and provide insightful comment, interpretation, and reflections on Gay's work: "Together, these elements allow the heat, grime, and smells, of the underbelly of 18th century London to come alive in new ways."

While a linear travelogue may comport to a known and mapped geography through which a story unfolds, the analogy of spatial storytelling and deep maps to a modern-day tour guide is worthy of consideration. Such a guide is often more than just a tourist map, for in its rich multimedia and multivocal format, the travel guide does not just map features and place names of a particular location or city. It also provides a richer description of a place that may contain information about cultural heritage, anecdotes, local personalities, shopping, entertainment, eating places, fictional works, and itineraries, along with facts and history about the place. The tour guide may even include multiple languages and a phrase book. All of this is told through a rich media mix of textual narrative, imagery, and maps. The tour guide may be the ultimate place-based spatial story and there is a reason that people take tour guide books on their travels and yet leave geography texts on their home bookshelves. The unfolding spatial story of a place revealed in numerous and interrelated forms are attractive in relating the experience of a place captured through a story guide of that place—a storied geography.

METHODOLOGICAL APPROACHES TO DEEP MAPS AND SPATIAL STORIES

But as Shakespeare's Hamlet would reflect: "Devoutly to be wish'd . . . perchance to dream: ay, there's the rub." Having identified images and visions of what deep maps and spatial storytelling might comprise, the question becomes how to convert such imaginaries and concepts into tangible and reproducible forms. Therein lies perhaps the greater challenge. Many approaches might be identified and proposed to achieve deep mapping or spatial narratives. An extensive investigation of these options is beyond the realms of this chapter but options and trajectories are necessary, otherwise this discussion is more one of optimistic wistfulness with little direction toward a means of implementation. I suggest nine immediate options and avenues perhaps worthy of pursuit.

First, the traditional hallmarks of intensive local historical investigation of cultural geography lies within the domain of historical geography. The work of Hoskins[42] in particular, with his focus on local history, archaeology, and historical geography, in many respects represents one of the very first examples of deep geography. Second, the map remains a powerful vehicle with which to interpret space and place, and especially so for storytelling. Examples from the Weave project where maps are linked to an explanatory voice-over are powerful forms of spatial storytelling.[43] The maps and accompanying explanation of elections in the United Kingdom, for example, convey a revealing and illuminating story to the reader. Kwan and Ding's[44] linkage of GIS, mixed-methods, and narrative forms provides a similarly revealing insight beyond the traditional spatial analytic forms of GIS. Third, ethnographic geographies combine much of the biographical storytelling embedded within everyday geographies. The work of Ghazi-Walid Falah,[45] for example, reconstructs the geographies of his childhood spent in Palestine to provide an experiential window into the exploration of memory, space, and attachment to place from the perspective of an indigenous inhabitant. Falah not only contributed to the making of place recounted through his experiences and childhood memories but he also subsequently used his story as a catalyst for political resistance to the forces of the state.

Fourth, the emerging field of qualitative GIS[46] provides both a conceptual base and a methodological platform to combine both the quantitative and qualitative traditions within geography and mapping. The theoretical base for qualitative GIS was born out of the GIS and Society debates of the 1990s and of critical GIS and critical cartography and the techniques were forged in the practice of PGIS.[47] In addition to the use of GIS, PGIS draws extensively on the local knowledge of communities in all the complex and qualitative forms that individuals and communities use to store and recall information and experiences. Qualitative GIS and PGIS with their emphasis on text, multiple media, representation, multiple truths, and mixed-method analysis may provide valuable options to explore deep mapping and spatial storytelling.

Fifth, a conceptualization of spatial stories and deep maps invariably involves some base in multimedia or geovisualization. An embedded media-rich digital environment would seem to be essential for much that is envisioned here in deep mapping. Spatial multimedia or multimedia GIS may be significant contributors to the implementation of these new forms of geographic expression. The power of virtual reality and the immersive and experiential capability provided by computer analysis and visualization environment systems would seem to be particularly applicable, and especially so when combined with the pursuit of a "sensual and reflexive GIS" that seeks to enhance the privileged visual with equal emphasis on sound, smell, and touch.[48] Sixth, and in a related vein, *The Wilderness Downtown*[49] is a particularly fascinating project that illustrates ways in which spatial science might be blended with other multiple media forms and personal experiences. In creating a real-time weave of location with moving images, virtual creations, aerial imagery, Google Streetscene, text, sound, and personal messaging in a creative and engaging style suggests ways in which new technologies such as HTML5 might provide exciting possibilities for deep mapping and spatial storytelling.

Seventh, I suggest that a storyboard environment for linking much of this material is critical to advancing the concepts of deep mapping and spatial storytelling. I have long advocated the use of the Prezi zooming presentation platform to be an excellent prototype for this approach.[50] ChronoZoom provides similar capability and possibilities.[51] The Prezi software enables objects, including text, imagery, video, and sound to be inserted within a presentation mode that has "infinite" depth. Within this storyboarding space, linkages and pathways can be established between and through the "presentation." While linear tracking is understandable given its primary presentation purpose, it does limit the types of woven stories that can be produced as alluded to previously. I embed this type of storyboard capability within a framework model provided at the end of this paper, but what is required is that consideration be given to the topology and greater freedom of navigation through the objects that are recorded and displayed.

Eighth, neogeography represents the confluence of geographic knowledge production and populist communication and interaction technology. New technologies premised on a socially networked Web 2.0 provide opportunities for a powerful geospatial web. The geospatial web represents a loose coupling of map mashups enabled through web mapping application programming interfaces. What initially appears to be a mishmash of concepts, data, and technologies can in reality contribute to a radical departure from traditional notions of spatial mapping. Geotagged social networking media from sites such as Flickr, YouTube, OpenStreetMap, Facebook, Wikis, and blogs can now be integrated with mapping platforms such as Google Earth, Google Maps, or Esri's MapsOnLine. The central characteristics of neogeography are threefold. First, "neogeography is about people using and creating their own maps, on their own terms and by combining elements of an existing toolset."[52] The creators and publishers of this data seek to create maps for themselves, the primary consumers of this data, using the building blocks of Web 2.0 technologies. Turner's second characteristic of neogeography is the user desire to share locational information with friends and visitors that helps shape context and understanding "through the knowledge of place." The emphasis on "sharing between equals" reinforces the vision of neogeography as a grass-roots, plebian, and popular movement. The focus on the layperson is in stark contrast to the expert, professional, specialized, and almost aristocratic nature of GIS as it has evolved to date. Finally, in contrast to the claimed lackluster nature of GIS, Turner suggests "neogeography is fun" because of its accessibility to the amateur and the use of mashups that give personal meaning and context to space and place.

Ninth, and connected to neogeography and, I would suggest a potential major contributor to the conceptual development of deep maps and spatial storytelling, is the role of citizen sensors and VGI as providers of personalized information about place. Using cheap handheld global positioning system–enabled devices, the layperson can now geotag most data with a digital spatial coordinate. Digital collection devices such as "smart" cell phones, digital cameras, and digital audio recorders now greatly ease the earlier difficulty of collecting, transferring, and uploading spatial data to the internet. Even the least technology-savvy person can now upload geotagged images to a social media sharing site for multiple viewing and download. Furthermore, these location-aware devices increasingly include critical metadata such as collection time and date and collection device, and keyword tags can be added automatically. Spatial information can also be tagged with latitude and longitude coordinates even without access to a global positioning system by using online resources such as address matching and sophisticated geocoders. All of these technological breakthroughs in social media technologies have occurred in rapid progression. The ability to upload media to social media websites such as Flickr and YouTube coupled with advances in cloud computing and storage have accelerated this trend since storage costs and media delivery are no longer a big obstacle to the average person. In addition, mapping these data through online mapping systems to create individualized mashups is equally accessible through the geospatial web. The cumulative impact of these user-generated content capabilities, coupled with spatial data services, social media technologies, and application programming interfaces–enabled online systems is that more local data is geotagged, stored online, and made retrievable by multiple users than at any other time in history.

To demonstrate how these elements might come together to form a deep map, I suggest a framework in figure 2.3 to illustrate how these vari-

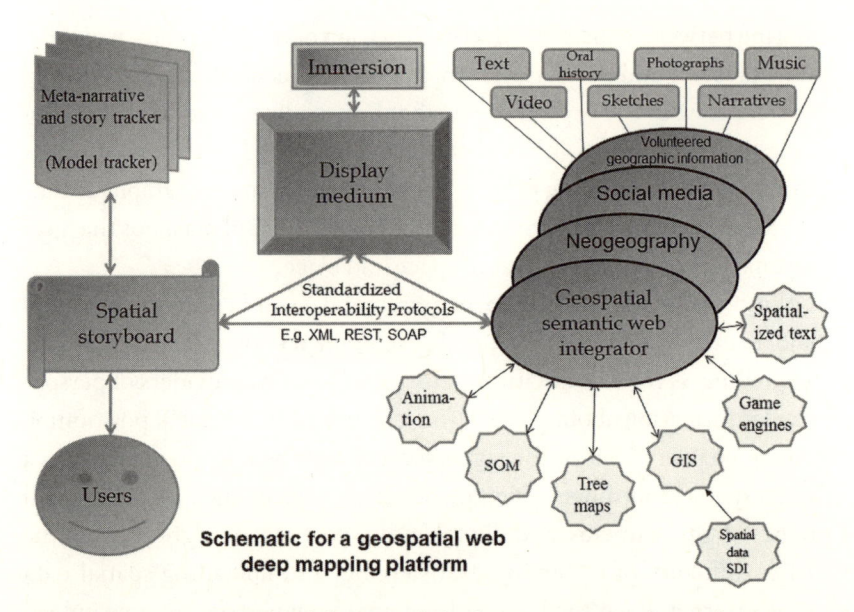

FIGURE 2.3. A geospatial web framework for deep mapping and spatial stories.

ous elements connect. The framework is premised on the power of the geospatial web to integrate disparate sources and multiple media based on now widely available application programming interfaces. Thus data might come from VGI, social media websites, authenticated spatial data, personal submissions, animations, images, text, tweets, and so forth, but all these can be combined though the integrative power of the geospatial web. A storyboard of infinite depth, such as Prezi, provides a medium through which to link the materials which can then be displayed on standard display devices or powerful immersive stereo-enabled systems such as a computer analysis and visualization environment. A more flexible topological system for developing differing linkages between the materials is critical. As noted previously, metadata is also critical to enable the tracking of material and the pathways taken, through the deep map and yet this is likely to remain one of the most challenging aspects of deep mapping. Neogeography and VGI represent a paradigm shift in spatial data generation and online mapping that collectively closes the gap between data producers and data consumers and provides a valuable framework in achieving a deep map.

PULLING IT ALL TOGETHER

Having laid out a conceptual base for deep maps and spatial storytelling and having identified possible methodologies, platforms, and approaches by which deep maps may be created, there remains a need to demonstrate how deep maps might be accessed and traversed to create spatial stories that resonate to specific users developing their own spatial themes. To this end, I suggest a metaphor to illustrate how such a weave of deep maps and spatial stories might take place. The Johnny Cash Project is a global collective art project where the theme is based on Cash's final studio recording of "Ain't No Grave."[53] The original video made of Cash recording this song was disassembled frame by frame and made available for download on the project website. Contributors were asked to select any one frame from this sequence and to render the drawing in any style they wished using a special drawing tool. These art styles range from pointillism to realism to sketch to abstract forms. The project grows as more people submit their drawings, which are integrated into a collective whole. The resulting art sequence, played to Cash's song, contains multiple frames of drawings of the video where each contribution is unique (figure 2.4). Listed below the timeline that tracks the music video are a series of boxes each of which represents a submitted drawing. In all instances, there are multiple submissions made for each frame. As the video-song is played, differing routes can be taken through the sequence of frames based on differing drawing genres, or by most recent submission, highest-rated submissions, or purely random tracks that generate differing combinations of frames. In this analogy, then, the totality of the frames represents the deep map where multiple "voices" have contributed media to the whole, in this instance a rendered frame, but, in reality, this could be any form of media. The story represents the sequence of frames traversed through the playing of the video based on some theme or objective such that no one pathway through the totality of frames need necessarily be the same for each journey through the sequence. Given the analogy I pose here for deep mapping and spatial storytelling, differing voices could be represented in the deep map through data, multimedia, and object frames. Thus multiple stories and multiple storylines can be tracked through the deep map and no one universal meta-narrative necessarily dominates. Within this structure, individual frames can be examined or themes and

FIGURE 2.4. Image taken from the Johnny Cash Project (http://www.thejohnnycashproject.com/) in which the multiple frames submitted can be replayed in differing ways through the playing of the video and song.

tracks selected which provide a deeper understanding of the media or knowledge and which guide the selection of frames. The system is an ever-evolving, ever-changing, dynamic set of contributions and tracked pathways through the deep map.

In the context of deep mapping and spatial stories, the Johnny Cash Project provides a powerful metaphor illustrating how deep maps and spatial stories may be constructed and accessed to provide a series of spatial stories through the deep map of user contributions. The deep map has the potential to be transformative in telling a story that is imbued with sound, images, maps, and a host of media that captures how people and organizations know about the world and represent that knowledge. In its original recorded form, the Johnny Cash Project tells a personal story, an emotional story. In its recast crowd-sourced form, it incorporates multiple reflections in the redrawn images that contribute richness and alternative understandings to the story. The ability to select themes through these series of contributed frames enables the story to be retold in differing ways and with potentially alternative perceptions and insight. Extending this metaphor to deep maps and storytelling provides one possible framework for imbuing deep maps and spatial stories with a powerful way in which to capture and represent multiple voices and ways of knowing.

NOTES

Note to epigraph: Warren (2004) cited on page 5 in G. Rambaldi, "Who Owns the Map Legend?" *URISA Journal* 17 (2005), 5–13.

1. William Least Heat-Moon, *PrairyErth: A Deep Map* (New York: Mariner Books, 1999).

2. B. H. Lopez, *Desert Notes: Reflections in the Eye of a Raven* (Riverside, N.J.: Andrews McMeel Publishers, 1976).

3. T. M. Harris, J. C. Corrigan, and D. Bodenhamer, "Challenges for the Spatial Humanities: Toward a Research Agenda," in D. Bodenhamer, J. Corrigan, and T. M. Harris, eds., *The Spatial Humanities: GIS and the Future of Humanities Scholarship* (Bloomington: Indiana University Press, 2010).

4. See M-J. Kraak and F. Ormeling, *Cartography: Visualization of Spatial Data* (Upper Saddle River, N.J.: Prentice Hall, 2002); M. Monmonier, *Mapping It Out* (Chicago: University of Chicago Press, 1993); A. H. Robinson, *Elements of Cartography* (New York: John Wiley & Sons, 1953); T. Slocum, *Thematic Cartography and Geographic Visualization* (Upper Saddle River, N.J.: Prentice Hall, 2003).

5. Robinson, *Elements of Cartography*.

6. Alan M. MacEachren, *Some Truth with Maps: A Primer on Symbolization & Design*. (University Park: The Pennsylvania State University, 1994).

7. Lopez, *Desert Notes*, 85.

8. W. G. Hoskins, *The Making of the English Landscape* (Leicester, U.K.: Penguin Books, 1955).

9. Heat-Moon, *PrairyErth*.

10. M. Pearson and M. Shanks, *Theater/archaeology* (New York: Routledge, 2001).

11. M. De Certeau, *The Practice of Everyday Life* (Berkeley: University of California Press, 1984).

12. Yi Fu Tuan, *Topophilia: A Study of Environmental Perception, Attitudes, and Values* (New York: Columbia University Press, 1974).

13. See K. Knabb, ed., *Situationist International: Anthology*. (Berkeley: Bureau of Public Secrets, 2006).

14. Knabb, *Situationist International*, 40.

15. G. Debord, *Introduction to a Critique of Urban Geography* (1955), reprinted in Knabb, *Situationist International*, 8.

16. Knabb, *Situationist International*.

17. T. M. Harris and H. Franklin Lafone, "Toward an Informal Spatial Data Infrastructure: Voluntary Geographic Information, Neogeography, and the Role of Citizen Sensors," in Kristyna Cerbova, ed., SDI, Communities, and Social Media (forthcoming).

18. T. M. Harris, D. Weiner, and T. Warner, "Pursuing Social Goals through Participatory GIS: Redressing South Africa's Historical Political Ecology," in J. Pickles, ed., Ground Truth: The Social Implications of Geographic Information Systems (New York: Guilford Press, 1995), 196–223; W. Craig, T. M. Harris, and D. Weiner, eds., Community Participation and Geographic Information Systems (London: Taylor and Francis, 2002).

19. W. Bright, Native American Place Names of the United States (Norman: University of Oklahoma Press, 2004); T. F. Thornton, "Anthropological Studies of Native American Place Naming," American Indian Quarterly 21, no. 2 (1997), 209–228.

20. Bright, Native American Place Names of the United States, 16.

21. Thornton, "Anthropological Studies of Native American Place Naming."

22. J. Harrington, "The Ethnogeography of the Tewa Indians," quoted in Bright, Native American Place Names of the United States.

23. See U.S. Board on Geographic Names at http://geonames.usgs.gov/antarctic/index.html (last accessed October 2013).

24. Pearson and Shanks, Theater/archaeology.

25. M. Pearson, In Comes I (Exeter, U.K.: University of Exeter Press, 2007); M. Pearson, Site-specific Performance (Basingstoke, U.K.: Palgrave Macmillan, 2010).

26. Pearson, In Comes I, 17.

27. J. Harvie, Staging the UK (Manchester, U.K.: Manchester University Press, New York: Palgrave New York, 2005).

28. Harvie, Staging the UK, 45.

29. Three Landscapes is available at http://metamedia.stanford.edu/~mshanks/three landscapes/index.html (last accessed October 2013).

30. A Map on a Wall (2001) is available at http://metamedia.stanford.edu/~mshanks/threelandscapes/map-on-a-wall.html (last accessed October 2013).

31. C. McLucas is quoted at http://metamedia.stanford.edu/~mshanks/projects/deep-mapping.html (last accessed October 2013).

32. P. J. Ethington, "Ghost Neighborhoods: Space, Time, and Alienation in Los Angeles," in M. Roth and C. Salas, eds., Looking for Los Angeles: Architecture, Film, Photography, and the Urban Landscape (Los Angeles: Getty Research Institute, 2001).

33. M. Twain, "Fenimore Cooper's Literary Offenses" (1895), http://twain.lib.virginia.edu/projects/rissetto/offense.html (last accessed October 2013).

34. P. Martin, How To Write Your Best Story—Advice for Writers on Spinning an Enchanting Tale (Milwaukee, Wis.: Crickhollow Books, Great Lake Literary, 2011).

35. See http://www.route40.net/page.asp?n=1 (last accessed October 2013).

36. M-P. Kwan and G. Ding, "Geo-narrative: Extending Geographic Information Systems for Narrative Analysis in Qualitative and Mixed-Method Research," The Professional Geographer 60, 4(2008), 443–465.

37. John Pickles, A History of Spaces: Cartographic Reason, Mapping, and the Geo-Coded World (Boca Raton, Fla.: Taylor & Francis, 2003).

38. Certeau, The Practice of Everyday Life.

39. J. Gay, Trivia: or, the art of walking the streets of London (1716 ed., Google Books), reprinted in C. Brant and S. E. Whyman, Walking the Streets of Eighteenth Century London (Oxford, U.K.: Oxford University Press, 2009).

40. John Gay, Book 1 in Brant and Whyman, *Walking the Streets of Eighteenth Century London*, 164.

41. Brant and Whyman, *Walking the Streets of Eighteenth Century London*, 1.

42. W. G. Hoskins, *The Making of the English Landscape.*

43. Information about Weave (Web-based Analysis and Visualization Environment) is available at http://oicweave.org/ (last accessed October 2013)

44. Kwan and Ding, "Geo-narrative."

45. G. Falah, "The 1948 Israeli-Palestinian War and Its Aftermath: The Transformation and De-signification of Palestine's Cultural Landscape," *Annals of the Association of American Geographers* 86, 2 (1996), 256–285.

46. M. Cope and S. Elwood, *Qualitative GIS: A Mixed Methods Approach* (Thousand Oaks, Sage Publications, 2010).

47. Harris, Weiner, and Warner, "Pursuing Social Goals through Participatory GIS"; Craig, Harris, and Weiner, *Community Participation and Geographic Information Systems.*

48. T. M. Harris, L. J. Rouse, and S. Bergeron, "Humanities GIS: Adding Place, Spatial Storytelling and Immersive Visualization into the Humanities," in M. Dear, J. Ketchum, S. Luria, and D. Richardson, eds., *Geohumanities: Art, History, Text at the Edge of Place* (New York: Routledge, 2011), 226–240.

49. *The Wilderness Downtown* is available at http://thewildernessdowntown.com (last accessed October 2013).

50. Prezi's website is http://prezi.com (last accessed October 2013).

51. ChronoZoom is available at http://eps.berkeley.edu/~saekow/chronozoom (last accessed October 2013).

52. A. Turner, *Introduction to Neogeography* (Sebastopol, Calif.: O'Reilly Media, Inc., 2006), 3.

53. The Johnny Cash Project is available at http://www.thejohnnycashproject.com/ (last accessed October 2013).

3

GENEALOGIES OF EMPLACEMENT

JOHN CORRIGAN

Keokuk, Iowa, calls itself "the geode capital of the world." For the last century and a half, collectors have walked the streambeds of the local tributaries of the Mississippi River extracting the popular rocks from the hundreds of outcroppings of the Lower Warsaw Formation that surround the city. Keokuk geodes come in many colors and sizes, but form in similar fashion, through seepage of silicates and carbonates into cavities within sedimentary rocks. The layering of sediments, hollowing of the rock, subsequent in-seepage of minerals, and eventual weathering of bedrock that results in exposure of the geodes has taken place over millions of years. The quartz formations that are found within many geodes, and that are treasured by collectors for their colors and brilliance, accordingly are the result of a process involving many factors: as strata are laid down over time, geodes are formed in new ways by the ongoing influence of groundwater and hydrothermal activity and surface as the material surrounding them decomposes.

The town of Keokuk is itself a place made by people who, as Michel de Certeau might say, have "toured" that geographic space for generations, laying down their own strata of practices of everyday life.[1] Over time, those strata have been buried and exposed and reburied, blended, altered and augmented, diminished, and articulated in ways that have defined Keokuk in a complex and contingent fashion. Keokuk, the place, like the hard, shiny quartz in a geode, can have the appearance of permanence, cultural heft, and solidity. At the same time, it clearly is a place constantly being remade, and particularly so in recent years as its economic base has suffered, its downtown made less familiar, and its crime rate turned sharply upward.

A deep map of Keokuk that brings forward the complex relationships among people, institutions, environment, culture, power, and the world outside Iowa begins with a concern for place and emplacement as defining aspects of the human. It begins with a gesture toward "homo geographicus" as one who makes place and is made by it.[2] The notion of "making" a place, as it has been developed by theorists such as Certeau, Michel Foucault, Henri Lefebvre, Edward Relph, Edward Soja, and others, is partly indebted to the philosophizing of Martin Heidegger, whose investigations into *Being and Time* (1927), as well as his National Socialist leanings in the 1930s, eventually led him to conceptualizations of place that subsequent writers have both embraced as pathbreaking and used as a foil in developing their own theoretical standpoints.[3] Heidegger's derivation of the word *dwelling* has been especially important in this regard. Heidegger associated dwelling with *building,* with its suggestions of constructive practice alongside the more crucial meaning of "to remain" or to "stay in one place." To dwell for Heidegger was to be at peace in a place, in the sense that one was free and at home there, a status that was marked equally by the naturalness of that and the fact that it unfolded in concert with a cognate project of preservation or care of the home. Such a theorization of place was less about individuals inhabiting a site than about engaging and preserving a "world." Heidegger explains it thus: "But if we listen to what language says in the word *bauen* we hear three things. 1. Building is really dwelling. 2. Dwelling is the manner in which mortals are on the earth. 3. Building as dwelling unfolds into the building that cultivates growing things and the building that erects buildings.... We do not dwell because we have built, but we build and have built because we dwell, that is, because we are dwellers."[4]

Whether we agree with the philosopher Jeff Marpas that Heidegger's work is "the most important and sustained inquiry into place to be found in the history of Western thought," it is arguably the case that humanistic geography has been profoundly shaped by Heidegger's insistence that any clarification of human experience of spatiality is possible only with a view to the spatiality of being in the world. Heidegger thus proposes:

> Space is split up into places. But this spatiality has its own unity by virtue of the worldlike totality of relevance of what is spatially at hand. The "surrounding world" does not arrange itself in a previously given space, but rather its specific worldliness articulates in its significance the relevant context of an actual totality of places circumspectly referred to each other. The actual world discovers the spatiality of space belonging to it. The fact that what is at hand can be encountered in its space of the surrounding world is ontically possible only because Dasein itself is "spatial" with regard to its being-in-the-world.[5]

In Heidegger's philosophy, place then was intertwined with human being, a position that he brought into relief by disapprovingly noting the thinking of his contemporaries: "For us today space is not determined by way of place. Rather, all places, as constellations of points, are determined by infinite space that is everywhere homogenous and nowhere distinctive."[6]

Heidegger's supporters and critics read his discussion of dwelling alongside his nationalism and his strong gestures toward regionalism in theorizing place. Some, like Edward Relph, found the notion of dwelling useful and sought to elaborate it in further theorizing space and place. Taking a phenomenological tack, Relph stressed that of "particular importance is 'existential' or 'lived' space," and by that he meant to convey that "space provides the context for places but derives its meaning from particular places."[7] For Relph, human consciousness was not disembodied and abstract, but, rather, an awareness of something "in its place," where place is understood to be objects and their meanings and a "concentration of our attentions" there. Places were not to be defined merely as locations or in terms of passing or superficial experiences of location, but, rather, as "profound centres of human existence." Sharing with phenomenologists such as Edmund Husserl and Merleau-Ponty an interest in the human as embodied subject, and holding that "places are thus the basic elements in the ordering of our experiences of the world," Relph summarized his thinking in his borrowing from the philosopher Gabriel Marcel: "An individual is not distinct from his place; he is that place."[8]

Like Relph, Yi Fu Tuan developed during the 1970s a view of place that carried forward the Heideggerian notions of "care" and "rest" (as well as "intentionality" and "essence"[9]) in the experience of place, and he coined the term "topophilia" to denote the "affective bond between people and place."[10] A place, layered over time with activities shared among a community of people, takes on the character of a "field of care" and evolves to "incarnate the experience and aspirations" of people.[11] Influenced as well by the cultural geographers J. B. Jackson and Carl Sauer, and uncomfortable with the materialist and functionalist trend that focused on the spatial patterning of economic, social, and political life, Tuan sought ways in which to relate the emotional and spiritual aspects of human life to space and place. The humanist geography that emerged in the 1970s, influenced directly and indirectly by his work, subsequently embraced a project of understanding "being-in-place," and as such adopted at least in part the essentialist, universalizing assumptions about humans and place that were redolent in Heidegger's work.

Theorists such as Relph and Tuan, among others who shaped a nascent humanistic geography, took their leads in part from their dissatisfaction with extremely quantitative research that left space as an empty Euclidean absolute. They also sought to complicate the regional geography of the time that focused on the distinctiveness of regions or areas. *Regional geography* as epitomized by the work of the German geographer Alfred Hettner (d. 1941) took the study of regions to be the central project of geography. As it developed in Europe and America, it focused variously on "ways of life" and general features of a particular region, leaving the profound engagement of the individual with space in general largely unanalyzed. Its method was typological and its aim was to define as ideal types the distinctiveness of regions, whether that meant nations, intercontinental territories (e.g., the Mediterranean), cities and towns, mountain country, and river valleys. Time, history, and the dynamic in many cases had a diminished position in such representations, and the attention given to the general characteristics—to clearly defined morphologies—of region impinged on representation of the individual and the unique.[12] This kind of chorology, which was best represented among Anglophone readerships by the work of Richard Hartshorne,[13] was not entirely deaf to culture, however. Regional geography as practiced by Hartshorne and Paul Vidal de la Blanche, and by Sauer—who was especially interested in historical change[14]—pushed off against environmental determinism[15] and sought ways in which to represent the interplay of people, culture, and setting that took into account environmental factors. The interpretations that resulted from that research tended to be idiosyncratic in that they focused on a single case in a particular time and place, as opposed to a nomothetic approach that aimed at generalizations drawn from a range of cases, times, and places.

Eventually, geographers such as David Harvey and Peter Haggett voiced criticism of the chorological approach that cast it as essentially descriptive and subjective and, as such, unscientific. Chorology, as J. Nicholas Entrikin has written, was inclined "to make selections and determine significance on the basis of personal judgment rather than tested generalizations, laws or theories." But as Entrikin has shown, regional geography could be both idiographic and naturalistic—in practice it relied upon both the subjectively managed discovery of cultural patternings and the insight that came with causal explanations that rested upon more generalized understandings of culture in space.[16] That regional geography was engaged in both of those kinds of inquiry—one roughly mimicking the humanities and the other naturalistic science—was in fact no surprise to Hartshorne himself, who wrote in 1939 (perhaps wryly): "Assuming only that geography is some kind of knowledge concerned with the earth, we will endeavor to discover exactly what kind of knowledge it is. Whether science or an art, or in what particular sense a science or an art, or both, are questions which we must face free of any value concepts or titles."[17]

The investigation of place and space in the latter part of the twentieth century developed a wide range of emphases, and art as well as science was present in branches of that research, although that was not apparent initially. The surge of interest in statistical and numerical methods in the 1960s–1970s resulted in pathbreaking syntheses, such as David Harvey's *Explanation in Geography* (1969) and Doreen Massey's *Spatial Divisions of Labor* (1984), that helped to clarify the scientific nature of the investigation of space at the same time that they disclosed the profound implication of the social in the spatial. Harvey called for a set of rules, grounded in scientific theory and logic, to shape geography as a truly explanatory scientific endeavor, and turn it from its inclination to poetic description. Deploying a Marxist perspective, he argued that "the problem of the proper conceptualization of space is resolved through human practice with respect to it." Accordingly, he concerned himself with "how distinctive human practices create and make use of distinctive conceptualizations of spaces,"[18] and how those spaces, as constructed by human practice, play their own role in framing social practices, fostering them, and challenging them. Massey, like Harvey, insisted upon theoretical sophistication and methodological precision in spatial analysis, and her work modeled the advantages of building analysis on clear conceptualization of space. She detailed the ways in which economic relationships and gender regimes, among other things, were spatially inscribed so that everyday practice was continuously shaped by that space.[19] Both writers, then, stressed the importance of scientific approaches to the study of space and made visible in research how lived space was defined by both everyday practice and by culturally established economic, gender, and other social ordering schemes. Both also, and especially Massey, rethought the study of region, and the latter concluded that place was characterized by processes of networks of social interactions, diverse identities and histories, and linkages to that which was "outside" the place. In defining place as such, Massey—among others who were thinking similarly in the late-twentieth century—recast regional geography as a focus on the dynamic, unique, and porous (in terms of relationships to other geographic spaces). That complication of place theory, which joined in several ways with Harvey's interests, raised the question of how geographic analysis as art might be implicated in its practice as a science.

The focus on the politics of space that informed the scholarship of Harvey, Massey, Nigel Thrift, Derek Gregory, Gillian Rose, and others, and that engendered fresh thinking about community and place, also prompted reconsideration of the poetics, alongside the politics, of geographical analysis and representation. Derek Gregory, who, along with Nigel Thrift[20] and David Ley[21] stressed in new ways the importance of actors' intentionality and agency in the layering of a place with structure and meaning, broached the potential of a geographical poetics for moving beyond a scientistic linguistics that had screened nuance, fluidity, and contingency and that in so doing had collaborated with a positivist agenda. "The form in which we render our accounts," wrote Gregory, "is not a secondary matter—a mere embellishment or ornamentation." The "crisis of representation" was one which "brought the politics of social inquiry and the poetics of social theory *into the same discursive space*."[22] Gregory pointed to Allan Pred's groundbreaking work on recovering the "lost words and lost worlds" of nineteenth-century Stockholm and Walter Benjamin's "chronicle" of his childhood in Berlin—both of which modeled the advantages of fashioning artistic language in renderings of place—as "the deliberate erasure of the finely etched line between the academic and the artistic."[23] The contingencies, contradictoriness, and multifacetedness of everyday life, the flows of power, and the constructedness of the *temperationes*[24]—the occluded social categories and rules that manage people's lives—all were more effectively represented in their particular contexts, through a more vibrant poetics. That point was reinforced by economic geographer Trevor Barnes who offered a comment on how a multiplicity of narratives had potential to disclose the layered and networked meanings of a place by responding to a critic, via a journal article, in five different ways.[25] Dennis Cosgrove, in writing about landscape, remarked on "the collapse of clear distinctions between science and poetics," and noted that "one expression of the recent change in geographical thinking is the shift from biological and cybernetic models of environmental and spatial organization to the organism or system, to metaphors derived from the arts, like spectacle, theatre, and text." He lamented that the writing of Renaissance cosmographers and humanists "has today been undermined by a deep, deconstructive distrust of aesthetics and poetics," and he called for a revaluation of their contributions and an openness to poetic language.[26] His thinking, as he pointed out, took up the position of John Wright, who, in 1947—about the same time that Hartshorne was rhetorically wondering about whether geography was an art or a science—had complained in the course of promoting "aesthetic imagining" in geography: "There are some who believe that we should explore only such *terrae incognitae* as lend themselves to exploration in accordance with rigorous scientific principles, that the purpose of such exploration should be to determine exactly what these *terrae incognitae* contain, and that in presenting the results to others we should aspire to strict, impersonal objectivity. It may be left, these say, to the artists, poets, philosophers, novelists, and politicians to develop the aesthetic and intuitive faculties of their minds; geographers should keep to a straighter and narrower path."[27]

When Michel de Certeau proposed that a place is made through the process of persons touring it, he was thinking in ways that resonated with some of these arguments about poetics as they were emerging in

Anglophone geographical literature. His thinking, however, developed its contours in ongoing conversation with other French writers, such as Michel Foucault, Pierre Bourdieu, Henri Lefebvre, and Bruno Latour—all of whom offered perspectives on space that similarly intersected with Anglophone geographical writing.[28] The spatial aspect of Foucault's thinking has been much discussed and debated, with Stewart Elden stressing that "Foucault's historical studies are spatial through and through," while critics such as Nigel Thrift have commented on the "blind spots" in Foucault's thinking about space.[29] Foucault located power in space and theorized the ways in which the spatial and social were intertwined. Power flowed through spatial fields by way of relays and capillaries and could be concentrated in institutions, such as the clinic or penitentiary. The movement of bodies in space proceeded as part of a process that included the ongoing renegotiation of the complex and multilayered ordering of society, the policing of social boundaries, the surveillance of individuals, and the heterogeneity and contradictoriness of lives. Foucault's attempt to spatially conceptualize the interrelation of individual bodies to those orders was through reference to the *heterotopia*, a lived site characterized by a sense of movement between the real and the unreal, or, simplistically put, as a process of ongoing remapping of identity and difference. According to Foucault, *heterotopias* as specific sites come into being for specific reasons and involve some measure of transgression of social categories and a sense of otherness, as in the case of a prison or a cemetery. For Foucault, not all places are *heterotopias*. If we were to try to define place in Foucauldian terms, we would stress both the durability and the fragility of orders and regimes, alongside flux, heterogeneity, the play between identity and difference and public and private, and the ongoing reconfiguration of the relationship between knowledge, power, and experience.[30] All of which is to say that place is more than material environment, structure, or individual experience.

The Foucauldian suggestion of place as neither structure nor actor represents an approach that is found in Bourdieu and Latour as well. Bourdieu's "theory of practice" was meant as an intervention into a gridlocked conversation about the relation between hegemonic structures and individual intentionality and initiative. Taking the body as a site of socially inscribed "dispositions" that shape taste, movement, judgment, sensibility, and feeling, Bourdieu proposed a "theory of practice" that emphasized their acquisition through everyday lived experience. *Habitus,* then, which refers to inscription so comprehensive and so deep that it is unnoticed by actors—like the air one breathes—is a spatial term that denotes a certain kind of place, one where structure and the subjective are joined in everyday practice. What is important for our purposes is that the concept of *habitus* presumes the relation of the body to space by means of practice that moves the body through space.[31] Latour, also calling our attention to practice, attempts to relate agency to structure through an emphasis on actors and the networks that they form. His notion of actors, however, is broad. Ontically equal actors (or "actants")—which can be human bodies, bottles, Wisconsin, medicine, the Black Sea, musical notes, a virus—form networks and then associations of networks in a way that, for Latour, should lead us to question the very notion of the social. Space as a result matters inasmuch as actors and networks come to be related one to another and in so doing mark space.[32]

The attention to practice, in its capability to create place, and the joining of that to a representational form, a specifically narrative form, is articulated in Certeau. In *The Practice of Everyday Life,* Certeau, inverting the usual terminology,[33] refers to place as a locative order of the "proper," where we find law, power, tradition—and "discipline" in the Foucaldian sense.[34] Space is imagined by Certeau as something like a network of antidiscipline, a field of tactics, where everyday activities continuously challenge, alter, and undermine disciplinary strategies. Place is associated with stability and univocity, while space is associated with practice: "Space is a practiced place. Thus the street geometrically defined by urban planning is transformed into a space by walkers."[35] The itineraries of those walkers form narratives. Space, unlike Heidegger's dwelling, is potentially unstable, more akin to Henri Lefebvre's conceptualization of "social space" and Edward Soja's notion of "thirdspace."[36] As bodies move through a place, they make stories. An itinerary, a "tour," is a narrative with a spatial syntax, in which an assortment of locations are linked and sites are joined, as space is "made by stories in the form of places put in linear or interlaced series."[37]

Writing by Certeau and others over recent decades has been concerned both with how persons live their lives in space and how interpreters can theoretically ground interpretation of those lives and the ways in which place is made. One way of looking at approach and method in the wake of these theoretical inquiries is to frame investigations of space and place as a kind of genealogy where that is understood as an investigation into the historically contingent layering of structures, strategies, tactics, discipline, antidiscipline, environment, everyday practice. Borrowing in a qualified way from Nietszche and Foucault, we might take a genealogy to be unconcerned with evidencing universals in history and uninterested in devising a univocal narrative of history.[38] A genealogy focuses instead on processes as they unfold in their local specificity and uniqueness, in interstices between everyday practice and structure. A *genealogy of emplacement* (GE) accordingly is an investigation into space and place that seeks to understand the ways in which place is made through everyday practice as that occurs in relation to structure. It observes a certain kind of "mixing" (*temperationes*) of practice with discipline and it seeks to identify the networks and the associations of networks that constitute narratives of place.

Genealogy of emplacement is a way of understanding place through analysis that takes seriously the hardening of strata into traditions, institutions, social orders, tastes, and habitual itineraries at the same time that it critically examines the immediate and everyday activities of actors in space. It recognizes that there is seeming permanence to place, at the very least a veneer of stability that is connoted by the Heideggerian dwelling.[39] It also notices constant alteration and instability. Genealogy of emplacement is an attempt toward clarifying the interplay of actors (or *actants*[40]) in networks and conglomerations of networks that give a place its distinctiveness. Its method is to narrate through curation, constructing complex collages in the form of deep maps. Its means of representation is the spatial narrative as that narrative emerges from the act of curation. If, as Certeau has suggested, place is made by stories, we define place not only by listening to those stories—engraved on space by those who travel it, and as they are stacked, eroded, and chipped, palimpsest-like—but equally by fashioning open-ended narratives about place that admit our own situatedness.

Curation as a means to understanding place can be undertaken under the broad umbrella of the digital humanities. Curation, specifically, can be practiced as a tactic of the spatial humanities, as that enterprise has arisen from the growing possibilities offered by geographic information systems and the enrichment of the overall project of the humanities itself that has been engendered by the "spatial turn." The nascent spatial humanities has much in common with the digital humanities, and especially in terms of research method. While recognizing that digital projects come in many shapes and sizes, with great variance from one to another as far as topic and approach, it is clear that in numerous cases there are commonalities. One of those commonalities is in the area of database construction and analysis, and especially the means and ends of that enterprise. Those who work with written texts, for example, have experimented with exploiting massive digital corpora through the algorithmic sifting of language. Searching enormous collections of texts—for example, the *Early American Imprints* or *Early English Books*—is a different kind of scholarly endeavor than closely reading a single text such as *Twelfth Night*. Armed with an algorithm built to search for words or word strings/arrays, computer-assisted textual research generates huge databases that can be examined for clues to the overall character of the collection. What distinguishes such research from standard literary analysis is that the digital scholar relies on the database that is created in this way to self-disclose, that is, to surprise the researcher with the patterns and correlations that emerge unexpected from each reorganization of the text. In short, the researcher does not approach the texts with a thesis about what will be found, and then look for ways in which that thesis can be corroborated or disproven. Rather, the researcher hopes to discover patternings that provoke new questions and redirect analysis into areas that have been overlooked. As digital theorist Stephen Ramsay has argued, such forays "seek not to constrain meaning but to guarantee its multiplicity."[41] In other words, there is open-endedness, not closure, and the overlaying of many narratives about the meaning of a collection of texts, or even a single text. But it is still the researcher who writes the algorithm, who in a very general way charts a path for the depiction of the texts as they are reorganized.

Curation is like algorithmic criticism. Curation involves the selection of artifacts and their arrangement relative to each other.[42] Curation as a method of the spatial humanities is the representation of artifacts in spatially defined environments. So, while the researcher in English literature might share with the spatial humanities researcher certain kinds of visualization strategies—the graph or pie chart, for example—to represent patternings in the data, the latter will explicitly invoke space as the master framework for determining relationships between artifacts, a strategy that the English literature researcher might not adopt. Those spatial relationships can be of many sorts, but the end result for the spatial humanities researcher will be a *deep map*. It will be a representation of space as populated by many varied artifacts that, taken together, narrate place, and in some instances, placelessness.[43] The cutting edge of the spatial humanities has concerned itself with geographic space and the making of place. Curation accordingly has been oriented toward the representation of items that can be located in geographic space. Some, such as county populations, city hall, railways and canals, police boats, commerce, baseball, coffee, or pollution are placed more easily than others, such as emotion, injustice, play, misogyny, or music. The representation of items and their locations in any event involves the building of a story about space.[44]

This brings us back to Certeau's notion of a person touring space—the city, specifically, for Certeau. We know space, Certeau implies, by moving through it, and by moving through it we make place (using those terms in the more standard fashion rather than Certeau's inversion of them). The irony in this way of looking at the role of everyday practice is that place, while taking on some of the features of Heidegger's dwelling, is at the same time more unstable than space. Place is in a certain sense more clearly defined than space, by the human activity than takes place there, while at the same time it is more dynamic. A deep map has to be able to capture that mix of stability and instability. The curator must populate the spatial representation with artifacts that allow for the visualization of networks of various sorts while at the same time leaving open-ended the story about exactly how those networks function. A GE should be expected to show patterns as well as randomness. There is the obvious emic/etic problem here: can a curator who is an outsider spatially arrange data in a way that does not fundamentally misrepresent the insider, the person who tours a city? Whose genealogy is it? There are possibilities for reducing the problem, and especially if a spatial humanities project proceeds on the assumption guiding much of the digital humanities that the referencing of a large mass of data—engaged at the outset without a prejudice (a "hypothesis") toward what it will reveal—can lead to surprises in its patternings once it has been embedded in a representational framework. For the spatial humanities, that representational framework is space. The narratives that we would expect to emerge from a GE, then, are not entirely free of investigators' biases, but they can be expected to emerge in ways that to some extent limit the reductions inherent in humanities discourses.

Finally, there is the matter of getting from theorizing about spatial narratives to creating them utilizing current technology. The simple answer is that we are not there yet. Scholars have been thoughtful about the possibilities, however, and have proposed a number of approaches ranging from illustrations of how qualitative data can be imported into geographic information systems research[45] to broad inquiries into the relation between curation and narrativity and the adaptation of technology to shape it.[46] For suggestive purposes—to point to rather than to prove—I reference Prezi, a free-use downloadable software that can be adapted to support modest deep mapping efforts.[47] What I am most interested in exploring using this software is the potential to employ it in a process of curation that is somewhat free-form while at the same time exploiting its features to give direction to the process of building a GE. We might consider this conception of the project to be analogous to designing an installation of art in a museum with many rooms. The rooms form pathways that guide persons in certain directions, to art that has been clustered or otherwise organized with some end in mind, while at the same time the exhibit allows for freer navigation of the space and consequently various engagements with the art. A spatial narrative can be a tour of a place that unfolds in ways planned by the digital scholar while at the same time allowing for alternate pathways through the space. The narrative that emerges from this kind of activity, then, is one that is constrained in certain ways by choices made by the digital humanist creator. It nevertheless offers opportunities for alternate "readings" and remains open-ended. It relies on the prospect of an emergent poetics of place that is grounded in a kind of triangulated relationship between the artifacts, the designer's arrangement of them, and the situatedness of the tourist.

Prezi, like some other programs, enables the imbedding of artifacts of many sorts—such as text, images, videos, charts, reports, audio, interviews—in whatever kinds of backgrounds the designer chooses, such as satellite, historical, topographic, or street maps, for example, or against other kinds of backgrounds. What is noteworthy about Prezi, however, is its easy capability to create pathways from one object to another. That is, a user can be actively guided through a place along predetermined paths. One might follow a person walking through the city, or driving through it, along a course where many different kinds of artifacts are engaged, having been placed on that path by the designer. The virtual tourist is free, however, to notice other artifacts along the way and to divert from the path to explore them, and in so doing alter the poetics of the tour and redirect the narrative about that place.

The full capability of a program like Prezi to function as a technological framework for a GE can only be realized in a large-scale project that offers a robust inventory of artifacts for the creation of a spatial narrative. But we can glimpse some possibilities. Turning our attention to Keokuk, Iowa, the geode capital of the world, we might ask how the people who live there tour the city differently than they once did, and from that query move on to other questions of how and why. We know that during the last thirty years or so, Keokuk welcomed a Walmart big-box store to a site outside of the downtown area, that eventually the store was closed and a church moved from downtown to that repurposed building, that a new Walmart Supercenter was built near the relocated church,[48] that the process of downtown decay sometimes associated with the arrival of Walmart has continued, and that crime has risen sharply, greatly exceeding national averages. People navigate the city in a way they did not a few decades ago, more often steering around the city center rather than through it, as can be seen in a Prezi presentation.[49] While place is being remade daily through the activities of persons outside downtown, the old core is losing personality and significance because of the diminishing and narrowing of human traffic that once marked it. The deterioration of place in this case is intertwined with the decay of built environment, and to a certain extent with more profound, related changes in topology and hydrology. In short, seemingly small developments can lead to profound consequences in the character of a place. Each layer has its crises and redefinitions that remain locked within the strata laid down over time in Keokuk.

NOTES

1. Michel de Certeau, "Walking in the City," in *The Practice of Everyday Life*, trans. Stephen Rendall (Berkeley: University of California Press, 1984), chap. 4, 91–110.

2. Robert Sack, *Homo Geographicus* (Baltimore, Md.: Johns Hopkins University Press, 1997). For a good overview of geographic thinking about place, see Tim Cresswell, *Place: A Short Introduction* (Oxford, U.K.: Blackwell, 2004).

3. See Stuart Elden, *Understanding Henri Lefebvre: Theory and the Possible* (New York: Continuum, 2004), 76–85 and *Mapping the Present: Heidegger, Foucault, and the Project of a Spatial History* (London: Continuum, 2001); Edward Relph, *Place and Placelessness* (London: Pion, 1976).

4. Martin Heidegger, *Poetry, Language, Thought*, trans. and intro. by Albert Hofstadter (New York: Harper & Row, 1971), 146.

5. Martin Heidegger, *Being and Time: A Translation of "Sein Und Zeit"*, trans. Joan Stambaugh (Albany: State University of New York Press, 1996), 97. See Edward Casey, *The Fate of Place: A Philosophical History* (Berkeley: University of California Press, 1998), 243–284.

6. Martin Heidegger, "On the Essence and Concept of φύσις in Aristotle's *Physics* B, 1"; in *Pathmarks*, ed. William McNeill (Cambridge, U.K.: Cambridge University Press, 1998), 190. Quoted in Jeff Marpas, *Heidegger's Topology* (Cambridge, Mass.: The MIT Press, 2006), 78.

7. Relph, *Place and Placelessness*, 8.

8. Ibid., 42–43.

9. Yi Fu Tuan, "Geography, Phenomenology, and the Study of Human Nature," *The Canadian Geographer* 15 (1971), 181.

10. Yi Fu Tuan, *Topophilia: A Study of Environmental Perception, Attitudes, and Values* (Englewood Cliffs, N.J.: Prentice Hall, 1974), 4.

11. Yi Fu Tuan, "Space and Place: A Humanistic Perspective" in C. Board, R. J. Chorley, P. Haggett, and D. R. Stoddard, *Progress in Geography*, 6 (1975), 236, and "Geography, Phenomenology, and the Study of Human Nature," 182.

12. Alfred Hettner, *Grundzüge der Länderkunde* (Leipzig, Germany: Teubner, 1923).

13. Richard Hartshorne, *The Nature of Geography: A Critical Survey of Current Thought in the Light of the Past* (Lancaster, Penn.: The Association, 1939).

14. On Sauer, see Ramesh Dutta Dikshit, *Geographical Thought: A Contextual History of Ideas* (New Delhi: Prentice Hall, 2006), 108ff., and Preston E. James, *All Possible Worlds: A History of Geographical Ideas* (Indianapolis: Odyssey, 1972)

15. Ellsworth Huntington, *The Pulse of Asia: A Journey in Central Asia Illustrating the Geographic Basis of History* (Boston: Houghton Mifflin, 1907).

16. J. Nicholas Entrikin discusses how there are there are both causal and descriptive sides to regional geographies that serve as a backdrop for humanistic geography in *The Betweenness of Place: Towards a Geography of Modernity* (Baltimore: Johns Hopkins University Press, 1991), 38–40.

17. Hartshorne, "Introduction," part C, para. 1.

18. Harvey, *Social Justice and the City* (Athens: University of Georgia Press, 2009), 13–14; originally published in 1973 by Johns Hopkins University Press.

19. Doreen Massey, *Space, Place, and Gender* (Cambridge, U.K.: Polity, 1994), 121, 269.

20. Some suggestive recent work by Nigel Thrift is "Cityscapes" in T. Beyes, S. Krempl, A. Deuflhard, eds., *Art and Urban Space* (Zurich: Verlag Niggli, 2009) 268–289; "Under-

standing the Affective Spaces of Political Performance" in L. Bondi, L. Cameron, J. Davidson, and M. Smith, eds., *Emotion, Place and Culture* (Aldershot, U.K.: Ashgate Press, 2009), 79–96; "Space, Time and Place" in R. Goodin, and C. Tilly, eds., *The Oxford Handbook of Contextual Political Analysis* (Oxford: Oxford University Press, 2006), 547–563.

21. David Ley, *The New Middle Class and the Making of the Central City* (Oxford: Oxford University Press, 1996).

22. Derek Gregory, "Interventions in the Historical Geography of Modernity: Social Theory, Spatiality, and the Politics of Representation," *Geografisker Annaler. Series B. Human Geography* 73 (1991), 20.

23. Allan Pred, *Lost Words and Lost Worlds: Modernity and the Language of Everyday Life in Nineteenth-century Stockholm* (Berkeley: University of California Press, 2005); Walter Benjamin, "Walter Benjamin, 'A Berlin Chronicle,'" in his *Reflections: Essays, Aphorisms, Autobiographical Writings*, ed. Peter Demetz (New York: Schocken Books, 1989) 3–60. "The poetics of my textual strategy are the politics of my textual strategy" (Pred, xv).

24. *Temperationes*, from Cicero, *De Natura Deorum*. I have proposed the word, which means to control, moderate, avoid, and especially "to mix in due proportion" to refer to those structurations of everyday life that guide thought, feeling, and behavior. I conceive of them as enforcing a cultural understanding of how individual lives "mix" with the life of the community, including regulations against certain behaviors as well as the fostering of others, and how those unique subjectivities are not lost in the social, but, rather, remain present in some measure as they are "mixed." *Temperationes* implies a ground upon which the creativity and spontaneity of individual practice is mixed with institutionalized power and imbedded ideology, but not in such a way as to completely subvert the authority or wholly undermine the effects of structure. When applied to discussions of place, it has something of the meaning of Michel Foucault's *heterotopias* in that *temperationes* suggest a "real" self, alongside an inverted or otherwise reshaped self (according to cultural rules), but *temperationes* does not imply a specific site where social orders or cultural commandments have been transgressed, as does *heterotopia*. *Temperationes* are present at all sites within a field of social relations. At some times the "mixing" takes place in ways that enable the continuance of the authority of, for example, class, gender, and feeling rules, while at other times the balance tips toward the alteration of those rules.

25. Trevor Barnes, "Five Ways To Leave Your Critic: A Sociological Scientific Experiment in Replying," *Environment and Planning A* 26 (1994), 1653–1658.

26. Dennis Cosgrove, "Environmental Thought and Action: Pre-modern and Postmodern," *Transactions of the Institute of British Geographers, New Series* 15 (1990), 345; "Globalism and Tolerance in Early Modern Geography," *Annals of the Association of American Geographers* 93 (2003), 854, 867.

27. John K. Wright, "The Place of Imagination in Geography," *Annals of the Association of American Geographers* 37 (1947), 6–7. The trajectory of "geographical imagination," part of which is marked by Wright's writing is sketched in Stephen Daniels, "Geographical Imagination," *Transactions of the Institute of British Geographers* 36 (2010), 182–187. A key text is David Lowenthal, "Geography, Experience, and Imagination: Towards a Geographical Epistemology," *Annals of the Association of American Geographers* 61 (1961), 241–260. Peter Fuller observed that "geography became stripped in its post-second-world-war years of its aesthetics and ethical dimensions. In the 1950s and throughout the 1960s, a positivistic 'scientism' began to fall, if not like a cloud, then at least like an exhalation of academic exhaust fumes across the discipline" ("The Geography of Mother Nature"

in *The Iconography of Landscape: Essays on Symbolic Representation, Design, and the Use of Past Environments*" ed. Dennis Cosgrove and Stephen Daniels [New York: Cambridge University Press, 1988], 12). The term is also central to Edward Said, *Orientalism: Western Conceptions of the Orient* (London: Penguin, 1978).

28. Michel de Certeau specifically addresses the perspectives of Foucault and Bourdieu in, *The Practice of Everyday Life*, chap. 4, 45–60.

29. John Marks summarizes the opinion of many that Foucault takes a "spatial approach to thought" itself in "A New Image of Thought," *New Formations* 25 (1995), 69, quoted in Andrew Thacker, *Moving through Modernity: Space and Geography in Modernism* (Manchester, U.K.: Manchester University Press, 2003), 22; Stewart Elden, *Mapping the Present*, 152; Nigel Thrift, "Overcome by Space: Reworking Foucault," in Jeremy Crampton and Stuart Elden, eds., *Space, Knowledge, and Power: Foucault and Geography* (London: Ashgate, 2007), 53–56.

30. Michel Foucault, "Of Other Spaces," *Diacritics* 16 (Spring 1986), 22–27. Thrift argues that because Foucault focused so much on space in terms of orders, he missed some of the vitality of space ("Overcome by Space," 55). But: "Power must be analyzed as something which circulates, or rather as something which only functions in the form of a chain. It is never localized here or there, never in anybody's hands, never appropriated as a commodity or piece of wealth. Power is employed and exercised through a net-like organization. And not only do individuals circulate between its threads; they are always in the position of simultaneously undergoing and exercising this power. In other words, individuals are the vehicles of power, not its points of application" (Foucault, *Power/Knowledge: Selected Interviews & Other Writings 1972–1977*, Trans. C. Gordon [New York: Pantheon Press. 1980], 98).

31. Pierre Bourdieu, *Outline of a Theory of Practice* (Cambridge: Cambridge University Press, 1977), and *The Logic of Practice* (Cambridge, U.K.: Polity, 1990); Jean Hillier and Emma Rooksby, eds., *Habitus: A Sense of Place* (Aldershot, U.K.: Ashgate, 2002); Tim Cresswell, "Bourdieu's Geographies: in Memoriam," *Environment and Planning D: Society and Space* 20 (2002), 379–382.

32. Bruno Latour, *The Pasteurization of France* (Cambridge, Mass.: Harvard University Press, 1988), and *We Have Never Been Modern* (London: Harvester Wheatsheaf, 1993). I am indebted to Graham Harman for his insightful discussion of Latour's metaphysics in *Prince of Networks: Bruno Latour and Metaphysics* (Melbourne: re.press, 2009).

33. Translations have rendered the French *lieu* as "place." "Location" is also a fair translation and perhaps a better one to avoid confusion.

34. Michel Foucault, *Discipline and Punish: The Birth of the Prison*, trans. Alan Sheridan (New York: Vintage, 1995). For Foucault, bodies are disciplined to maximize their docility under the surveillance of a panoptic power and discipline is constituted as a "unitary technique by which the body is reduced as a 'political' force at the least cost and maximized as a useful force" (221).

35. Certeau, *The Practice of Everyday Life*, 117. As such, space, unlike Heidegger's dwelling, is potentially unstable.

36. Henri Lefebvre, *The Production of Space*, trans. Donald Nicholson-Smith (Oxford: Blackwell, 1991); Edward Soja, *Thirdspace: Journeys to Los Angeles and Other Real-and-Imagined Places* (Cambridge, Mass.: Blackwell, 1996).

37. Certeau, *The Practice of Everyday Life*, 115.

38. Genealogy is understood here less in Nietzschean fashion than in terms of Foucault's foregrounding of historical contingencies and downplaying the predictably rational in the ordering of knowledge/power.

39. Although all is changing, but on different temporal scales, some literally in geologic time, others from moment to moment.

40. "Actant" in the sense suggested by Latour as nonhuman and inorganic as well as human.

41. Stephen Ramsay, "Toward an Algorithmic Criticism," *Literary and Linguistic Computing* 18 (2003), 167.

42. A recent comment on the digital humanities averred that it "recasts the scholar as curator." "Digital Humanities Manifesto," 2009, http://www.humanitiesblast.com/manifesto/Manifesto_V2.pdf (accessed May 13, 2014).

43. Relph, *Place and Placelessness*, 79–121.

44. In a venture that resonates in certain ways with some current thinking about visualization strategies for the spatial humanities, the DECIPHER project has begun exploring ways to build narratives from museum artifacts. See http://www.decipher-research.eu/ (accessed May 13, 2014).

45. Mei-Po Kwan and Guoxiang Ding, "Geo-Narrative: Extending Geographic Information Systems for Narrative Analysis in Qualitative and Mixed-Method Research," *The Professional Geographer* 60 (2008), 443–465.

46. See Relph, *Place and Placelessness*, 79–121.

47. See http://prezi.com/index/ (accessed May 13, 2014). Thanks to Trevor Harris for suggesting this software.

48. Discussion about specifically religious aspects of place has been influenced by the theories of Mircea Eliade and Jonathan Z. Smith, among others, most of which are inclined toward Certeau's notion of "poetic geography" (*The Practice of Everyday Life*, 105).

49. Thanks to Adam Brasich for collaborating on the project. See http://prezi.com/ahuyvbfht142/first-christian-church-disciples-of-christ-keokuk-iowa/ (accessed May 13, 2014).

4

INSCRIBING THE PAST

Depth as Narrative in Historical Spacetime

PHILIP J. ETHINGTON AND NOBUKO TOYOSAWA

In the humanistic disciplines, maps have a historical dimension. The historical dimension of any map is, in one leading sense, its "depth." This is the focus that we shall pursue, through an examination of examples from seventeenth-century Japan and twenty-first-century United States. What we mean by *deep map* is one that is historically deep. Its historical depth gives it a narrative dimension. Deep mapping and spatial narrative, therefore, are essentially interwoven, as we shall argue in this essay.

The humanities—studies of human being and meaning—are intrinsically historical and geographic. Every past always has a place. Human actions *took place* and *made* all of the places that we know today. To tell stories about the past is to tell stories about places made by previous generations. Making places of the past has involved both physical and textual actions. Place-making is also accomplished by telling stories about the relation between those actions in places of the past and our present. Actions in writing today make and remake the past places of the Earth. Further, writing about places and actions constructs those places and actions in collective memory. *Reading* maps and textual narratives also makes and remakes the places of the past. Mapping the textual past is therefore an interactive archeology (a metaphoric archeology in Foucault's sense). In such an archaeology, we read and write layers of meanings that reference all of the embodied places in so many literary *topoi*, or textual "places" that serve as references of semantic meaning.[1]

MAP 4.1. Los Angeles Basin showing municipal boundaries with dates of incorporation, aerospace production sites, boundaries of the 1965 Watts Riot, and Uto-Aztecan village sites, circa 0 CE–1769. (Original in color). Cartography © 2011 Philip J. Ethington.

In this essay, our primary objective is to explicate and provide examples of deep mapping as narrative: *in* the past, *of* the past, and *about* the past. The past of any place is a narrative, as we show with examples of place-making in centuries past. Intellectuals in the past narrated the past of their nations, a process of place-making and history-making that is inseparable. Our own practice as historians is a part of this tradition, which we argue is the necessary awareness of every historian—that the past is a specific set of places on Earth and in texts.

"Mapping" should be construed very widely. Maps represent the relationships among the elements of any kind of topography—those of a terrestrial landscape, or of a metaphoric "landscape" of texts, ideas, or networks of relations among people. The *Oxford English Dictionary*'s definitions of the noun *map* and the verb *to map* range from "the representation of the earth's surface or part of it on a flat surface," to the "conceptualization or mental representation of the structure, extent, or layout of an area of experience, field of study, ideology, etc."[2] Thus we can map the streets of Dublin, or we can map the poetic structure of James Joyce's *Dubliners*.

Maps can be "of" or "about" natural features, the built environment, demographics, actions, events, experiences, and ideas. Since Edward Tolman introduced the term *cognitive map* in 1948, and the urban planner Kevin Lynch coined the term *mental map* in 1960, these terms have been applied broadly to encompass not only an individual's mental representation of their surrounding geometric spatial environment, but also the ways that individuals imagine all of their environments: those of work, residence, of travel, of their nation, and of the global world.[3] Henri Lefebvre developed the concept that spaces are "perceived," "conceived," and "lived," and that the way people perceive, conceive, and live in space is historically conditioned.[4] Indeed, the history of cartography has convincingly demonstrated that there is no certain or universal representational mapping of any part of the universe. All pictorial maps are extreme abstractions of the physical world, highly inflected with biases of inclusion, omission, scale, and projection. The same characterization can easily be made of textual, semantic maps. Any "map" of James Joyce's *Dubliners,* or of the influence of Joyce in modernist literature, will be highly subject to idiosyncratic influences of the mapmaker. In short, mapping is a creative and culturally embedded practice.[5]

No matter what the subject of a map is, it can take many forms. Pictorial maps are what most people mean by *maps:* cartographic representations of surface features of the Earth, with textual annotations, primarily consisting of place names, or *toponyms. Textual maps,* in this essay, are defined as nonpictorial "mappings" of some spatial relations conducted via natural language, without reliance on pictorial representation. In short, textual maps take a narrative and descriptive form. The Early Modern European term for this practice is *chorography:* extended descriptions of places, including some of the following—although not always in the same combination: environmental features, human inhabitants, landmarks, productive elements, and cultural and political history.

Pictorial mapping and textual mapping are two very different methods of representation. Cartography as a pictorial form operates by *simultaneity* and *juxtaposition*; semantic text is syntactically linear and narratological. As such, each form of communication can do something that the other cannot. Neither is superior, and both are complementary to the other.

A *spatial imaginary* is the mental image of any spatial environment, held by participants in a symbolic discourse. We can infer the spatial imaginary of a society from the ensemble of its pictorial and textual maps, as found in manuscript records of older societies, print circulation of more recent societies, and the mass media and internet communications of contemporary societies. The following sections of this essay map a history of place-making as a process of inscribing spatial imaginaries into the landscape. In it, we hope to model a method of deep mapping that can account for the present as a topography of *inscribed places*.

JAPANESE *FUDOKI* WRITING FROM THE EIGHTH TO TENTH CENTURIES

In the case of Japan, the close connection between the past and places dates to the time of the Nara period (710–794). In 713, Empress Genmei (660–721, r. 707–715) issued an edict ordering the local provinces to report about the matters in each of their districts to the Yamato court, the first ruling body to exercise authority over most of the Japanese archipelago. The edict specifically instructed local officials to submit regional gazetteers that expressed the names of the provinces, districts, and villages in good characters; explain the origins of the names of mountains, rivers, plains, and fields; list special products of the province (excluding agricultural products); describe the fertility of the earth for agriculture purposes; and report strange and old hearsay tales that were handed down from generation to generation by the elders.[6] Each local province reported back to the court, and the reports were compiled into sixty-five volumes and presented to the court later during the reign of Emperor Daigo (885–930, r. 897–930).

These documents came to be called *fudoki* (literally, the records of the wind and earth, more commonly referred to as the regional gazetteers).[7] They explain the genesis of place names—what the place was originally called and how the place name was replaced with different characters, or why such changes occurred in the first place, while also recording the special products of the regions and their geographic features.[8] The wealth of these topographic details included in *fudoki* testifies to the historical depth of all the places in the provinces. The edict was successfully implemented because local officials enthusiastically embraced it by displaying the richness of their own local provinces.[9]

One example of the change in place names is found in the *fudoki* of the Harima province (*Harima no kuni fudoki*), which is the southwestern part of today's Hyōgo Prefecture. It discusses the change of a place-name "Shino" (死野—field of death) to "Ikuno" (生野—field of life). People had come to call the place Shino, the "field of death," because there lived an angry deity, Kōjin, who killed people who passed the area, so the field was filled with a sense of death. According to the *Harima no kuni fudoki*, upon hearing the deeds of the angry deity, a quasi-historical emperor in the late fourth century, Emperor Ōjin, also known as *Honda no sumeramikoto*, revised the name to Ikuno to transform the place into positive-spirited space.[10] This is one example of layered histories inscribed in places, drawn from *fudoki*—the oldest documents in Japan to record histories that took place in the local provinces, including such mythohistorical events.

In *fudoki*, there are generally sections devoted to listing the origins of local shrines and temples, as well as the names of villages and districts. As with the example of Shino/Ikuno, the sections explain how the particular place came to be called by its name. It is noteworthy that they often have connections to deities or to emperors—some of whom were believed to have descended from the deities.[11] Some of these figures are historically controversial, to be sure. Regardless, *fudoki* records make it clear that many places in Japan refer to the creation stories of the archipelago. *Fudoki* assiduously recorded such local ties with the deities that are inscribed in the place's name. Looking into the names of mountains, rivers, plains, or fields reveals deep histories that are embedded in these places. Some stories that overlapped with the myths concerning the origin of the Japanese archipelago are recorded in Japan's oldest chronicles of the eighth century, *Kojiki* (Record of Ancient Matters) and *Nihon shoki* (Chronicles of Japan). The indigenous practice of deity worship for the protection of the local communities from epidemics, natural disasters, and other misfortunes, or, conversely, in wishing for good harvest and good fortune, is reflected in the changing names of places in Japan.

THE REVIVAL OF THE *FUDOKI*-STYLE WRITING IN THE SEVENTEENTH CENTURY

In the late seventeenth and early eighteenth centuries, as ever more Japanese took to the road, a Confucian scholar and traveler, Kaibara Ekiken (1630–1714) of Chikuzen Province (present-day Fukuoka Prefecture), developed a new mode of writing about space and topography by measuring the traveling distances and investigating topographic features that corresponded to descriptions in ancient and medieval texts.[12] Ekiken wrote about various local places based on a variety of mythohistorical texts and worked to invigorate a sense of historical continuity of the ancient past during the Tokugawa regime of "peace under Heaven."[13] Ekiken's writings are reminiscent of the ancient *fudoki*-style, which he recognized very explicitly:

> The topographic writings (*chishi*), written about the "wind" and "earth" of our land (*honchō no fūdo no koto*), are ancient documents issued during the reign of Empress Genmei. The empress, for the first time, ordered the five Home Provinces and seven Circuits to make records about each province and district in the region.... After generations, Emperor Daigo [897–930] in the sixth year of Enchō (928) completed the compilation of regional gazetteers with sixty-five volumes that recorded details of the sixty-six provinces. Indeed, they are truly important documents for our realm. Later, the Meitoku Rebellion [in 1391] and the Ōnin Rebellion [1467–1477] saw the destruction of these records. Only the *fudoki* of Izumo and Bungo provinces are extant, although not in full records. Alas, how unfortunate the loss is, indeed![14]

Urging the need to resume the reproduction of *fudoki*, Ekiken lists contemporary scholars who revived the tradition of *fudoki* and *fudoki*-style writing.[15] His own work is evidently dedicated to the purpose reinventing the *fudoki* tradition.

The compilation of national history became the central occupation for Ekiken and other intellectuals from the seventeenth century onward.[16] They used Chinese, the universal scholarly language in East Asia, to write Japan's history, while often modeling their work after Chinese historiography and scholarship. Ekiken applied the Confucian evidentiary method to Japan's local places, investigating their validity through official historical texts.

While he based his writings on the *fudoki* tradition, and sought to revive it, Ekiken did so in the new genres of the rising print culture of the Tokugawa era. Ekiken's writings were published as tour guides, illustrated display books, and reference lists that circulated new spatial imaginaries for the emerging reading public.[17] He published his research and investigations of these places in the form of tour guides to the capital of Kyoto, to the ancient capitals of Nara and other cities, to a renowned hot spring Arima, and to Shogun Ieyasu's burial ground Nikkō. Ekiken's spatial narratives attracted a publisher in Kyoto, the scholarly and publishing center of Japan, who published them as guidebooks in the late seventeenth and early eighteenth centuries, and even after Ekiken's death.[18] Mainly based upon historical and topographic investigations that he conducted during his own travels, these tour guides consisted of a series of short descriptions, with few illustrations and almost no pictorial maps. To understand the structure and meaning of Ekiken's spatial narratives, we will focus now on a textual analysis of *A Record of the Arima Hot Spring* (*Arima no yama onsenki*, 1711). Ekiken narrated the history of the Arima region by extracting and remapping the significance of its places.

The structure of *A Record of the Arima Hot Spring* is composed of eight major sections: (1) the way from Kyoto to the Arima hot spring, (2) some wisdom about the hot spring treatment for illnesses, (3) historical information about the Arima Mountain, (4) other renowned hot springs in the country, (5) mountains and rivers in the Arima area, (6) special products of Arima, (7) different routes from (and to) Arima, and (8) poets

MAP 4.2. Ekiken's route to Arima from Kyoto. (Original in color). The map displays the general path taken by Ekiken from the Tōji Temple in Kyoto to the final destination of the Arima hot spring in the guidebook, *Arima no yama onsenki*. The labeled place markers represent every site mentioned in the guidebook, a list of places introduced as places worthy to view. It is apparent from the presentation of sites that the reader was assumed to travel generally along the main roads, with no specific path recommended between every site in the guidebook. Instead, the reader was encouraged to explore north and south of the general direction of travel. Thus, the sites are arranged in an east-to-west order, but not strictly so. Map data © 2012 Google Earth, ZENRIN, GeoEye, DigitalGlobe. Published in conformity with Google Earth Terms of Service (2014).

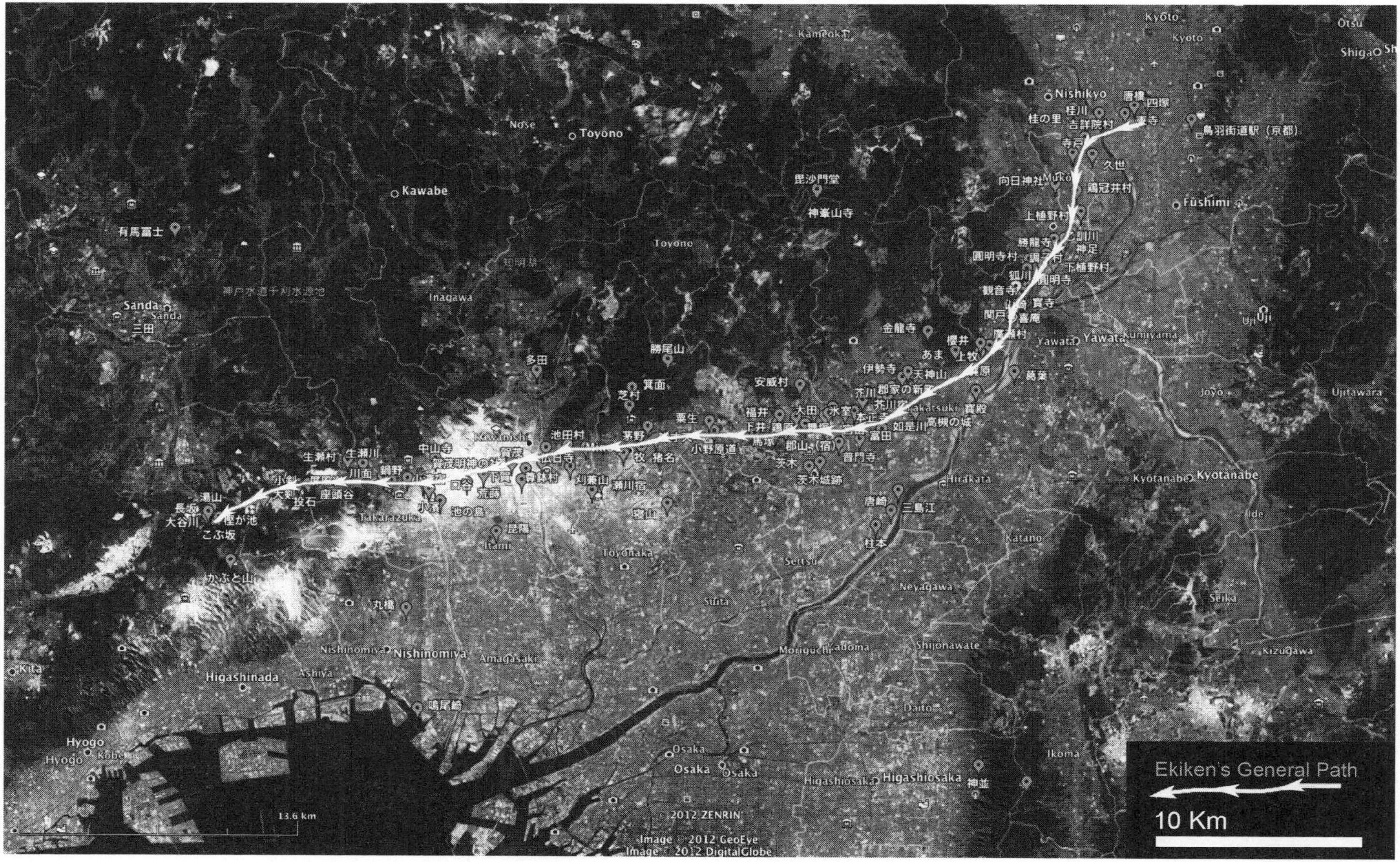

Ekiken's General Path
10 Km
Kyoto
Otsu
Shiga
Kameoka
Nishikyo
唐橋
四塚
東寺
桂川
桂の里
吉祥院村
鳥羽街道駅（京都）
寺戸
久世
向日神社
Muko
鶏冠井村
Fushimi
上植野村
勝龍寺
神足
下植野村
圓明寺村
狐川
圓明寺
観音寺
関戸
宝寺
喜庵
毘沙門堂
神峯山寺
金龍寺
櫻井
上牧
あま
天神山
伊勢寺
安威村
郡家の新家
芥川
Takatsuki
高槻の城
如是川
眞殿
富田
葛葉
Uji
Yawata
Kumiyama
Joyo
Ujitawara
Kyotanabe
Ide
Hirakata
Katano
唐崎
三島江
柱本
福井
大田
氷室
下井
馬塚
郡山（宿）
茨木
茨木城跡
音門寺
粟生
小野原道
茅野
牧
猪名
勝尾山
箕面
芝村
多田
池田村
Kawanishi
中山寺
瀬川宿
刈乗山
豊鉢村
荒蒔
池の鳥
昆陽
Itami
寝山
Toyonaka
Settsu
Suita
Neyagawa
Moriguchi
Kadoma
Shijonawate
Daito
Ikoma
Seika
Kizugawa
Higashiosaka
神並
生瀬村
生瀬川
川面
鍋野
座頭谷
投石
湯山
長坂
大谷川
こぶ坂
かぶと山
Takarazuka
丸橋
鳴尾崎
Nishinomiya
Amagasaki
Ashiya
Higashinada
Hyogo
Kobe
Kita
Osaka
Sanda
三田
有馬富士
Kawabe
Toyono
Nose
Inagawa
13.6 km
© 2012 ZENRIN
Image © 2012 GeoEye
Image © 2012 DigitalGlobe

who visited Arima and composed poems about Arima. This structure is reminiscent of the ancient *fudoki* edict that demanded local provinces to report about histories of the names of their mountains and rivers, along with their special products.

Arima is one of the countless hot springs found in various parts of Japan whose history lies deep in ancient time. Ekiken's tour, however, contains no justification for choosing Arima over other famous hot springs. Ekiken provides a historical overview of Arima by quoting the *fudoki* of Settsu province—a province located in the eastern part of today's Hyōgo Prefecture and the northern part of Osaka,—the eighth-century record of the *Collection of Myriad Leaves* (*Man'yōshū*), poems compiled in famous imperial anthologies, and the *Chronicles of Japan*. The *Chronicles* records the visit of Empress Jomei (593–641, r. 629–641) to Arima, which establishes that the history of the Arima hot spring dates to the seventh century, more than a thousand years prior to Ekiken's writing. Emperor Jomei was not the only ruler to have visited Arima; as Ekiken briefly mentions, Emperor Kōtoku (596–654, r. 645–654) did so, and, much later, Toyotomi Hideyoshi (1537–1598)—one of the unifiers of the warring states era—loved the Arima hot spring and invested money to build bridges, inns, and bathhouses. This historical continuum is literally marked in the landscape of Arima in the form of monuments, sacred rocks, springs, and other objects recognized by inhabitants as signifying the actions of the past.

The tour to the Arima hot spring starts from the Tōji Temple in the middle of Kyoto and continues in the southwesterly direction until the destination of the Arima hot spring. Structurally, the arrangement of sites is dictated by east-to-west order of travel, which determines the development of Ekiken's narrative. However, the order does not provide the narrative elements of emplotment, such as building up the traveler's emotions or the implicit reader's excitement toward starting the trip to the age-old hot spring, except in the pedestrian sense of what sites are proximal to a traveler as they proceed along a road. From the start to the end of this one-way trip, the traveling distance is fourteen *ri* (about thirty-four miles), covering about one hundred suggested stops, which include places that became famous due to their poetic association, their liveliness as being a port or river, spiritual power, and scenic beauty.[19] In other words, this is a practical tour guide, describing each place purely with factual information. For example, a place known as Kōnabi is described as follows:

> Kōnabi (神並): This is where the deities lined up, so the name of the place is deities (神) lining up (並). This is also a *meisho* [literally "place of name," but carrying the sense of "place of fame"; these places became famous from repeated mention in the Japanese poems called *waka*] and there are many poems about it. In the provinces of Yamato and Tango, there is a place by the same name [with different characters] 神南, but they are pronounced "Kaminami."[20]

Then, Ekiken introduces the next few stops, such as Kajihara, Tanba Valley, the Ise Temple, and others, while pointing out a rare kind of cherry blossom called *mazakura*, strange-looking rocks, different temples and shrines, castles of the local lords, and steep mountains in the background. The narration develops simply as the tour progresses from one place to the next, nearing the destination, and on first reading, at least, there seems to be no particular reasons or justification why Ekiken has chosen these stops in the tour over other places that are also towns, temples, and shrines.

And yet, there is a hidden narrative that is voiced through the chain of selected places in the tour, if we apply an interpretive lens. The starting point of the tour, the Tōji Temple, was the center of the Esoteric Buddhism, which was an international form of Mahayana Buddhism that had a strong emphasis on individual salvation founded in Japan by Kūkai (774–835).[21] Kūkai opened his monastic center on Mt. Kōya, which became the sacred forest where he meditated and trained himself and eventually his disciples. The Tōji Temple was Kūkai's headquarters in Kyoto where he disseminated his teachings. We argue that reading Ekiken's guide with these spiritual connections, his tour begins to illuminate the magical power of the holy ascetics associated with Esoteric Buddhism, and thereby takes the readers into the deep historicity of this mountainous region of Arima.

An excerpt from Ekiken's narrative from a place called Ōda to the next several stops allows us to examine his method. The stops included are within close proximity, as shown in map 4.3.

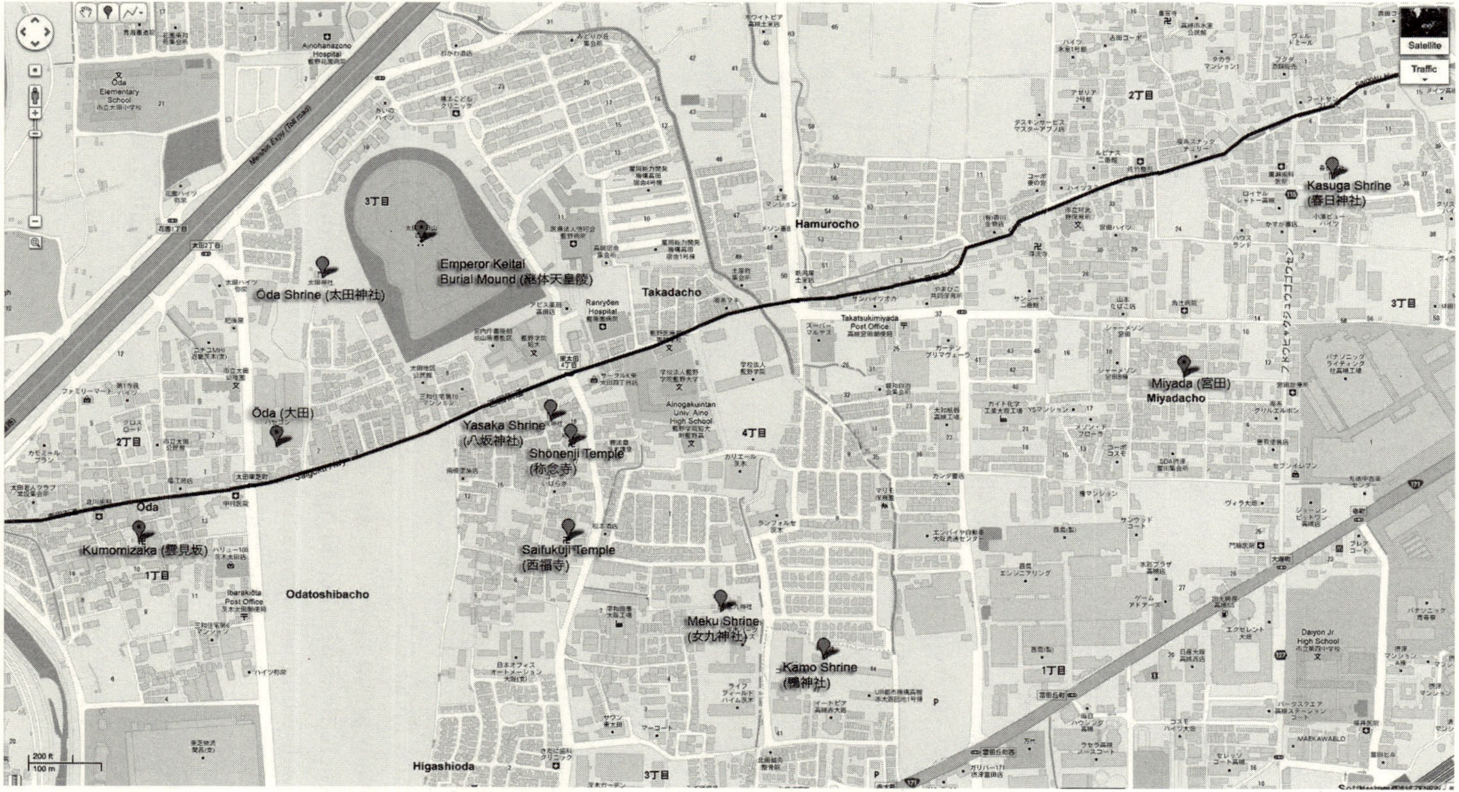
Satellite
Traffic
Kasuga Shrine
(春日神社)
Hamurocho
Emperor Keitai
Burial Mound (継体天皇陵)
Oda Shrine (太田神社)
Takadacho
Takatsukimiyada
Post Office
Miyada (宮田)
Miyadacho
Oda (太田)
Yasaka Shrine
(八坂神社)
Shonenji Temple
(称念寺)
Ainogakuintan
Univ. Aino
High School
Oda
Kumomizaka (雲見坂)
Saifukuji Temple
(西福寺)
Ibarakioda
Post Office
Odatoshibacho
Meku Shrine
(女九神社)
Kamo Shrine
(鴨神社)
Daiyon Jr.
High School
Ainohanazono
Hospital
Oda
Elementary
School
Ranryōen
Hospital
Higashioda
200 ft
100 m

ōda: There is a village. This is where people bleach cotton cloth [in the river]. There seems to be a small mountain, but it is an imperial tomb for Emperor Keitai [r. 507–531], and the tomb is known as *Aino no misasagi* (imperial mound of Aino field). This tomb is recorded in *Engishiki*. He is buried far away from his capital, *Yamato no tamaho* [or *Iware no tamaho no miya*, which is believed to be located in Nara.]

ai village: is located to the right. To the west, there is the Ai Mountain. Originally here is where a tomb of [Fujiwara no Kamatari (614–669) who held the rank of] *"Taishokukan"* [also known as *Taishokan*, meaning the great woven crown, which was the highest rank in a political system used only between 647 and 685. The first and only person ever granted the rank was Kamatari, the most powerful courtier of the seventh century] was located, but later it was moved to the Tōnomine Mountain in the Yamato province. There is still the ruin of the tomb. The mountain is not too high.

tōkaichi village: From here to the east, there is a path toward the Sōjiji Temple. From here to the Sōjiji Temple is about one *ri*.

sōjiji temple: This is the twenty-second *fudasho*, sacred site for pilgrims [where amulets might be offered]. There is about one and half *ri* from Akutagawa [introduced as a place of inns shortly before]. The temple is located in a high place in the village of Sōjiji. There is a tower gate, and this is a Shingon Buddhism temple.

minoura: [no description]

map 4.3. Midway to Arima on the Saigoku Road (Original in color). To the Arima hot spring, Ekiken took the Saigoku Road, one of the roads organized by the Tokugawa shogunate, which took people from Kyoto to Shimonoseki, a city located on the western tip of the main island. The map displays the Saigoku Road as a heavy black line (now lying between two major expressways to the north and south). Surrounding the Saigoku Road, Ekiken introduced several places worthy of visit. The four actual places mentioned by Ekiken are marked by place-markers with a black dot: Miyada, the burial mound for Emperor Keitai, Ōda, and Kumomizaka slope. The place markers without dots are temples and shrines that are located near the path but are never mentioned by Ekiken. The Meku Shrine is, for example, said to have memorialized the wives of Emperor Keitai (r. 507–531) upon his death. The Ōda Shrine is registered in the tenth-century *Engishiki*. Map data © 2013 Google, zenrin. Published in conformity with Google Maps Terms of Service (2014).

NUKAZUKA: Is located to the right of the road. It is a round mound, also known as Haizuka. One theory is that this is the imperial mound for Emperor Keitai.

TEKURA RIVER: Is located to the south of Nukazuka. The forest of Tekura (*Tekura no mori*) is a *meisho,* and the poem about *Tekura no mori* is compiled in [the thirteenth-century poetry text by Emperor Juntoku [1197–1242, r. 1210–1221] *Yakumo mishō* (Revered Notes on the Art of the Eightfold Clouds).

FUKUI: Is located to the right.

SHIMOI: Is the same as Fukui.

MUMATSUKA: Is located on the left.[22]

Ekiken moves without explanation from one place to the next, passing by numerous alternative sites. While each of his chosen stops is located in the close proximity of no more than one *ri,* many other choices were certainly available to him. A crucial question, then, is why did he choose the places he did? The east-to-west narrative structure of the guide is the first consideration, and, in particular, the mountainous terrain of Arima and the spiritual landscapes in it are important elements to decode the meanings attributed to this tour.

Ekiken carefully provides the readers with directions, right or left, north, south, west, or east for them to easily navigate the tour. The descriptions are telegraphic: entries for each place include only factual information, such as the core identity of the place being a *meisho,* what sect of Buddhist schools of thought, topographic details, or to which emperor the imperial mound belongs. Ekiken's narration works as a *narrative map* for the readers. What we mean by this term is that, as with most of his publications, the Arima guide offers only a few illustrations of the landscape, and no pictorial maps. In spite of the lack of pictorial maps, the readers can orient themselves spatially by following Ekiken's narration. Punctuated by the directions and traveling distances with minimum information about the places, the readers are aptly provided not only to follow his tour but also to visualize the spaces in the deep mountainous Arima region as vividly as if the reader were in the landscape. If we think of travel along the surface of the landscape as "horizontal," and the imaginary travel backward in time as "vertical," then readers can orient themselves horizontally as they travel mentally or physically from one place to the next.

At the same time, as they read through Ekiken's descriptions, the readers can travel vertically, learning and feeling the historical depth of Ekiken's carefully selected places by identifying the moments in the past through the present. Thus, Ekiken creates a spatial narrative of Arima that is joined by linking the horizontal places with the vertical historical depth.

Close inspection of this deep-referenced spatial narrative yields even more clues. Within the traveling distances of fourteen *ri* from the start of the Tōji Temple to the last stop, Ekiken often selected stops with place-names related to the deities, such as Kōnabi (神並) "the deities are lined up," Kōdani (神足), "a deity's legs," Kanpūsan (神峯山), "the god's peak mountain," and *Sae no motomura* (道祖の本村), referring to "the original village of dōso tutelary deities of roads and border." We argue, therefore, the most probable way to read his narrative was to assume that Ekiken was tracking the ancient practice of naming local places after the acts or manner of the deities. He does not necessarily inform the readers with every detail about the naming of the places, but the inclusion of these places with the character *kami* (神, deity) signals Ekiken's attempt to narrate history through local landscapes in which deities reputedly resided. Indeed, the mountain region of Rokkō, where the Arima hot spring is located, is a region saturated with connections to mythical stories, consisting not only of deities but legendary human heroes.

For example, Ekiken briefly introduces the historical account of the Shinjuji Temple (*Shinjuji engi*), informing the reader that this is where the legendary Empress Jingū buried the six suits of golden armor, along with the arrows, bows, and swords, upon returning from her conquest of Korea in the second century. The six suits of armor are symbolized by the iron helmet (*kabuto*), so the name of the mountain (山) became six (六) warrior's helmets (甲), Mt. Rokkō (六甲山) or simply helmet mountain (甲山). Empress Jingū's additional military supplies were said to be buried in the Mt. Mukoyama (武庫山), literally meaning, the mountain (山) of the warehouse (庫) of weapons (武).[23] The space where the Arima hot spring is located is thickly associated with the stories of this legendary warrior empress.

As Ekiken's tour continues, he describes the surrounding environment of the Arima hot spring, noting three tutelary shrines, namely, *Yu no gongen, Miwa myōjin,* and *Kashita myōjin,* all located to the south of a hall enshrining the Medicine Buddha (*Yakushidō*). He also describes another such shrine on top of the mountain, which reveres *Atago gongen*.[24] These *gongen* and *myōjin* are Japan's indigenous deities manifested as Buddhist deities. It was believed that they are indigenous deities, who, after the arrival of Buddhism, now appeared in the form of Buddhist deities to guide the Japanese for salvation just as Buddha led numerous ordinary people for salvation.[25] Ekiken discusses these indigenous deities similarly with the Zen sect temples, such as Onsenji and the Rannyain Temple, along with a Sōtō sect Zenpukuji Temple, which are all located in the mountainous surrounding landscape of the Arima hot spring.[26]

In sum, Ekiken's narration has the effect of encouraging the reader to imagine the natural landscape of the mountainous region as overlapped with sacred spirituality. As the narration continues to include more sacred sites, we can further glean the underlying significance of his choices. Included are three stops along the traditional Buddhist pilgrimage of thirty-three temples in Western Japan (*Saigoku 33-sho junrei*): the twenty-second Sōjiji Temple, the twenty-third Kachiodera Temple, and the twenty-fourth Nakayamadera Temple. Ekiken does not explain the ways each of these temples are dedicated to *kannon*, the Japanese name for the Bodhisattva of Compassion or the Goddess of Mercy. Neither does he illuminate what the traditional pilgrimage is all about.[27] By identifying these temples of the pilgrimage of the Goddess of Mercy (*junrei kannon*), which are all located in the mountain known as *reizan* (spiritual mountains), Ekiken inscribes a certain kind of spiritual aspect to the sites. Very subtly yet symbolically, Ekiken's tour consists of a highly significant subset of temples and shrines, including those that enshrine fusions of Buddhism and indigenous deity worship. These sacred sites selected by Ekiken were not exclusive to the royal family and the court nobles but, rather, were constructed for the benefit of ordinary people.

The connection to mythical figures, such as Empress Jingū, or to Buddhist temples, does not mean that the Arima region is, or was, inherently sacred. Rather, these stories inscribe or reinscribe sacredness into the place, and Ekiken's narration of Arima through both symbolic and physical connections with places of deities further infuses these places with spirituality. More precisely, Ekiken selects places that have ties with mysticism to ascribe the sanctity into the place, simultaneously reviving the indigenous idea that Japan was a country of the deities. In fact, Ekiken's place-narrative reflects his faith that Japan was a country descended from the deities. In *Fusō kishō* (Record of Scenic Japan), an unpublished essay that Ekiken wrote later in his life, he identified Japan as a birthplace of life. In this text, such qualities as honesty and fairness of the people are equated with the just nature of gods. This idea of Japan being a country of deities (*shinkoku*) is repetitive in Ekiken's other writings, also expressed in another unpublished work, titled *Jingikun* (Precepts of the Deities of Heaven and Earth). *Jingikun* is his theorization of the historical origin and the nature of the way of deities (*shintō*). Ekiken explains that the prosperity of Japan is deeply rooted in the celebration of the deities of heaven and earth and articulates reasons for respecting the deities of heaven and earth, how to revere them, and the relationship among deities, Confucian sages, and Buddha.[28] For example, using similar language to that in *Fusō kishō*, Ekiken describes Japan's origin as:

> Our country Japan ["*waga hi no moto*," which means, our country where the sun rises first] is a *shinkoku* (country of deities) from the time immemorial, and the country created by the deities. Therefore, the nature of the people, their behaviors, and customs are superior to those of various other countries. Because Japanese are superior to others, Japan is said to be more sagely (*kunshikoku*) than China, people are wealthy and happy, their etiquettes are courteous, and that Japan is different from other countries. Even now, this trend from the ancient past has not changed. People are offspring of the deities, and whether they are of noble or low birth, everyone is equipped with the principle of deities (*shinri*) at heart.[29]

Ekiken repeats this logic of Japan being a country of deities throughout the text, claiming, based on the mythical origin of the country, that the nature of Japanese people is just and honest. He goes on to claim that because the country was created by the deities and ruled by them, the name *shinkoku* has been Japan's name since the ancient past, just as naturally as China is called the country of the Sages (*seijinkoku*).

Through this contextual analysis of his other writings, we can see that Ekiken's conviction about Japan as "*shinkoku*," country of the deities, is reflected in his intellectual pursuit to identify the names of places and their connections to the deities. He openly criticizes the practice of some Japanese rulers who wished to be deified in their afterlife by building *hachiman* shrines and *tenjin* shrines, and turning themselves into some kind of buddha.[30] This is an implicit criticism of Tokugawa Ieyasu and Toyotomi Hideyoshi, who were respectively enshrined posthumously at the Nikkō Tōshōgū Shrine and Toyokuni Shrine. Rather, what Ekiken seemed to encourage—at the time when the Tokugawa shogunate came to enjoy the peaceful realm called the realm under heaven—was to revive the ancient practice of revering the deities of the heaven and earth.

Observed in this light, Ekiken reinscribed histories of Japan into local landscapes, and his place-based history produced an ideological formulation of Japan as a country of deities. On the one hand, Ekiken's dense and often dry writing might not have attracted a popular readership and was more or less used as a reference book. The significance of his work remains the same because the places marked in Ekiken's guidebooks became representative sites for the local provinces and the "famous places" that were worthy of visit. There are various implications of Ekiken's deep mapping of mythical deities and spiritual sites. For one, his claim runs parallel to his contemporary scholars who challenged the centrality of the Chinese Middle Kingdom and pursued a renewed place of Japan within East Asia. With the revival of the *fudoki*-style writing, which was a way to legitimize the court authority, the Tokugawa scholars were certainly beginning to decentralize China by first reproducing the histories of the country. Their attention to the local places confirms our claim that the act of writing histories is inseparable from places because all "matters of our wind and earth" did *take place*.

SEMANTIC AND VISUAL MAPPING

Kaibara Ekiken's deep maps of Japanese historical landscapes demonstrate some universal qualities of deep mapping. First, the landscapes onto which Ekiken projected his *fudoki*-style narrative of historical pasts are *not uniformly mapped*. He selected only a small subset of possible sites, leaving many other temples and shrines, as well as most of the actual lived region, unmapped. Maps of any form, be they pictorial or textual, only spottily represent the total lived landscape in any era.[31] If the lived world were really like the maps we make of it, we would fall into huge gaps of nothingness between small islands of reality. Maps redact and condense, throwing out most of the available mappable terrain. Second, be they fictional or scientific, semantic *textual narratives thread a very narrow path* through the lived environments so scantily portrayed. Whether we consider Homer's *Odyssey,* Melville's *Moby-Dick,* Chaucer's *Canterbury Tales,* or Ekiken's *Arima no yama onsenki,* each cuts a thin path through space-time, leaving most of the world unnoticed and undisturbed by human representation. Third, *texts are inscribed into places.* Ekiken's *Arima no yama onsenki,* shows how semantic (textual) meaning becomes literally attached to unique geophysical locations on Earth. As he takes the reader through the region between Kyoto and Arima, Ekiken verifies the poetic claims made about places. He attests that they were the places of enactment of previous deeds, or that they are featured in a poem in the literary canon. His textual claim becomes thereafter a landmark, or *topos.* Cultural actors inscribe meaning into places in all times, and sometimes intellectuals attempt to map some of that meaning. They often produce new meaning about places, which then accumulates until others build and transform those meanings. Fourth, *meaning is projected from one place to another* in historical time, in effect stretching the cultural signification from one homeland to another. Ekiken explicated the religious and spiritual significance of Japanese places, reinscribing a Chinese influence that had begun one thousand years earlier.

It can be said, then, of all lived places of the globe, that each has a deep historical past. This is to say that they have deep inscriptions of remembered meaning from many generations. Factually, this is why all places are unique. Each is a mix of the influences that were inscribed by previous generations. Even the humblest village is the work of many centuries. Urban places have served as intersections for the long-running trade between many cultures and ways of life. Ethington's "ghost maps" of the Los Angeles Basin's thirteen-thousand-year human history present our next example of deep mapping of the inscribed past.[32]

GHOST MAPS: NARRATIVE DEEP MAPPING OF LOS ANGELES

We next offer ghost maps as an example of deep mapping as a pictorial form of historical representation that offers a strategy of recounting the past of a place, which is very unlike Ekiken's purely textual histories of places. And yet, ghost maps are also about the same kinds of things as Ekiken's are. They tell stories about a place, assuming that the reader decodes their pictorial narrative. *Ghost maps* are so named because they make visible the invisible traces of past human action in the landscape. As a rich and complex graphical composition drawn directly from the profound complexity of past social life itself, the content of a ghost map—like the inhabited urban environment itself—exceeds the capacity of textual narratives that could "explain" or interpret it. Ghost maps were designed to be stand-alone pictorial maps with minimal text: only labels and legends. They can be used for interpretation, study, and reference, and they can be part of a larger textual project to frame the narratives of a place. Ghost maps of the past are certainly complemented and magnified by textual maps. So, we argue that the combination can provide a much more robust representation of the global past of every human society in any era.

Map 4.4 is a small reproduction of Ethington's 2010 ghost map, "Geveronga–Yaanga/El Pueblo de Nuestra Senor Reina de Los Angeles/La Placita/Downtown/Los Angeles, California, USA 0 CE–2000 CE." The title itself is a palimpsest of place names for the region of Los Angeles, California, that is generally known as the "Downtown" of that great global metropolis. The area represented in this ghost map contained several hundred thousand inhabitants in the year 2000. In that year it was the center of a vast metropolis of about seventeen million persons. The complex history of this place, inhabited continuously for more than twelve thousand years, is very great and very deep. The ghost map's title declares a set of past eras that are inscribed into its places that it represents: covering more than two thousand years and four regimes of societies who ruled the Los Angeles region (Uto-Aztecan, Spanish, Mexican, United States).[33]

> Hic, ubi nunc Roma est, orbis caput, arbor et herbae et paucae pecudes et casa rara fuit.
>
> *Here, where now stands Rome, capital of the world, there were trees, and grass, a few sheep, and occasional cottages.*
>
> —Ovid, *Fasti* (8 CE) book v, lines 93–94[34]

Twenty-first century Los Angeles is the product of more than one hundred centuries of human effort that transformed the deep ecology of the Los Angeles Basin into one of the world's greatest nodes of global power and cultural interaction. Networks of roads, freeways, rails, ports, and

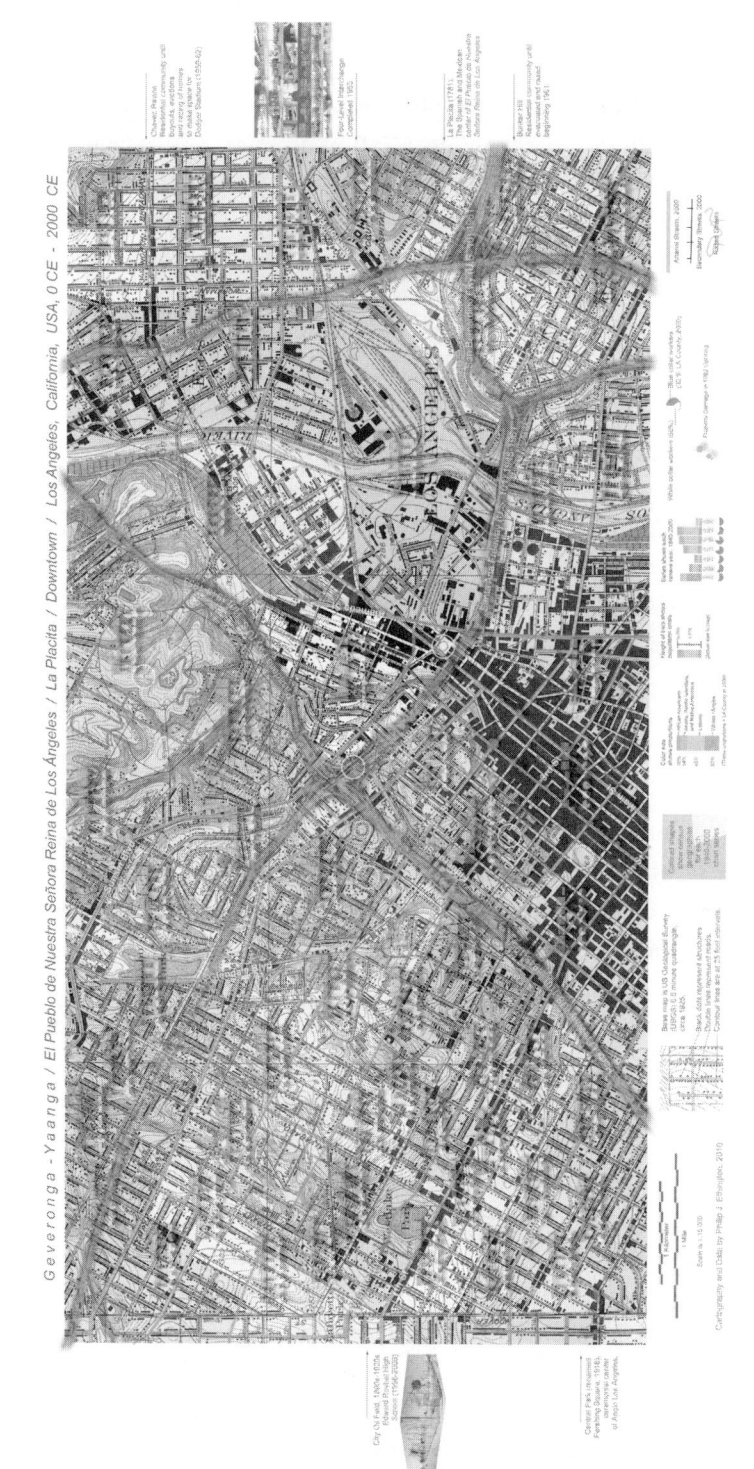

MAP 4.4. Ghost map (Ethington, 2010), "Geveronga—Yaanga. El Pueblo de Nuestra Senor Reina de Los Angeles/La Placita/Downtown/Los Angeles, California, USA 0 CE–2000 CE" (Original in color). Cartography © 2010 Philip J. Ethington.

airports draw the labor of about eighteen million persons (in 2009), and dynamically integrate a mighty concentration of power and capital that was developed over the last few centuries around the site of an ancient settlement called Yaanga.[35] Los Angeles is the expression of successive regimes of power established by Uto-Aztecans, Spanish Mexicans, and Americans from the United States who exploited an ancient alluvial basin and the people within it; the resulting riches have attracted and sustained migrants from every continent.[36]

Ghost maps of these histories are pictorial maps: works primarily of visual representation of lived historical landscapes. Ghost maps are designed visually to reveal time, change, events, and motion through the symbolic languages of color, shape, iconography, and textual annotation. They are *not* made to be read quickly, as mere illustrations or diagrams. Rather, each ghost map is a free-standing document, offered for the reader to ponder and to puzzle over, to return to many times.

The pictorial narration in map 4.4 begins with the earliest known place-names: those of the Uto-Aztecan villages that Spanish conquerors encountered when they invaded this region in 1769 with intent to take control of it for the Spanish Empire. By our best estimates, these villages were established about two thousand years ago by the Uto-Aztecan conquerors of a territory held prior to that by a culture known only as the Millingstone People. It is presumed that the Uto-Aztecan invaders seized the best dwelling sites, then already thousands of years old, from the Millingstone People, but it is not known what the Millingstone People called them.

Place-naming is place-claiming. Naming a place is the most elemental form of claiming it for a society. In Uto-Aztecan society, each village of about two hundred to five hundred people was a complete social entity; there was no "tribe" or "nation." Uto-Aztecans were a hierarchical, patrilineal society with a permanent, hereditary aristocracy, a middle class, and a slave class. So deeply did these Uto-Aztecans inscribe their place with meaning as a collectivity, that the headman of each village, the *Tomyaar*, assumed the name of the village as his own during his period of rule. In a sense, the *Tomyaar* was inscribed with the same toponym as the place. And each of these "triblets," as the anthropologist Alfred Kroeber called them, had its own collective identity. While we call them Uto-Aztecans today after their language and cultural descent group, they only knew themselves separately as members of perhaps forty such villages in the Los Angeles Basin.

When the Spanish arrived, they proceeded to strip the region clean of indigenous signifiers. They renamed the Uto-Aztecan peoples, formerly known to themselves only by their village names, such as Geveronga and Yaanga, with a new signifier: "Gabrielinos," meaning those associated with the Misión San Gabriel Archangel, the church and enforced labor hacienda established to convert the indigenous people into Christians and Spanish subjects. And the Spanish also massively renamed the topography of the basin they claimed. The central settlement, built atop the former Yaanga, was "christened" (declared to be a Christian space) with the long and cumbrous title "El Pueblo de Nuestra Señora, Reina de Los Ángeles del Río Porciuncula." In English: "The town of Our Lady, Queen of the Angels, on the River Porciuncula." The River Porciuncula runs through the birthplace of St. Francis of Asisi, the founder of the Franciscan order to which the church leaders of the Spanish conquest of California belonged.

The base layer of the ghost map is a U.S. Geological Survey topographic map from 1924. It shows the topography of the still-young metropolis, by clearly indicating Spanish and Mexican Rancho territories. The 1997 street lines show the imprint of many new neighborhoods constructed since the 1920s. The Uto-Aztecan village site of Yaanga, the Spanish and Mexican Placita, Anglo Downtown, and twentieth-century dense settlement are all represented. White circles placed over key locations where events took place are indexed to annotations in the margins of the map. These two-dimensional call-outs are another method of deepening the map. Further, these annotations lead to additional visualizations of the historical past of the locations annotated. Figure 4.1, "Ghost Neighborhoods" is a photomontage of a razed city block in 1997, with 1939 watercolors of houses that once stood on that block, superimposed as transparent images.

The ghost map also simulates "time depth" by the accumulation of transparent layers above the ancient-to-more-recent landscape traced by the topographic lines of the 1920s U.S. Geological Survey map. Horizontal time series in pie and bar charts animate the changes in class and ethnic composition of each neighborhood, from 1940 to 2000, while visual depth is accomplished with semitransparent layers that mark time vertically from lower (deeper past) to higher (more recent). There are many

FIGURE 4.1. Ghost Neighborhood: *Court and Beaudry, 1939 and 1997* (1997). (Original in color). Image © 1997 Philip J. Ethington.

more stories than there is space to tell them, legible in this ghost map. Each neighborhood has a separate story. The deepest are in La Placita, the circled plaza in the upper right of the detail in map 4.5. Aligned southwest from the Placita are the *calles principales* (main streets) of the eighteenth-century Spanish layout. These are represented as solid black by the U.S. Geological Survey symbology because by the 1920s they were built-up solid with 120-foot concrete and brick office structures along the Anglo city's central business district. The open square in the lower left of map 4.5, Pershing Square, is the Anglo recentering of the city in the industrial U.S. era. (General Pershing invaded Mexico in the Punitive Expedition in 1916, before leading U.S. Expeditionary Forces to France in 1917–1918).

Innumerable deeds could be narrated as having taken place in the spaces represented by the ghost map. The ghost map is not limited to any of them, and yet can reference any of them by sites: apartments, restaurants, streets, squares, street railways, and so on. The ghost map has many stories to tell in its multiple narrative series, dimensions, and iconographic representations. But narrative stories of people's lives, plots, intrigues, deaths, and dramas, are told off-camera, as it were, from this pictorial representation of the spaces in which those actions literally *took place*.

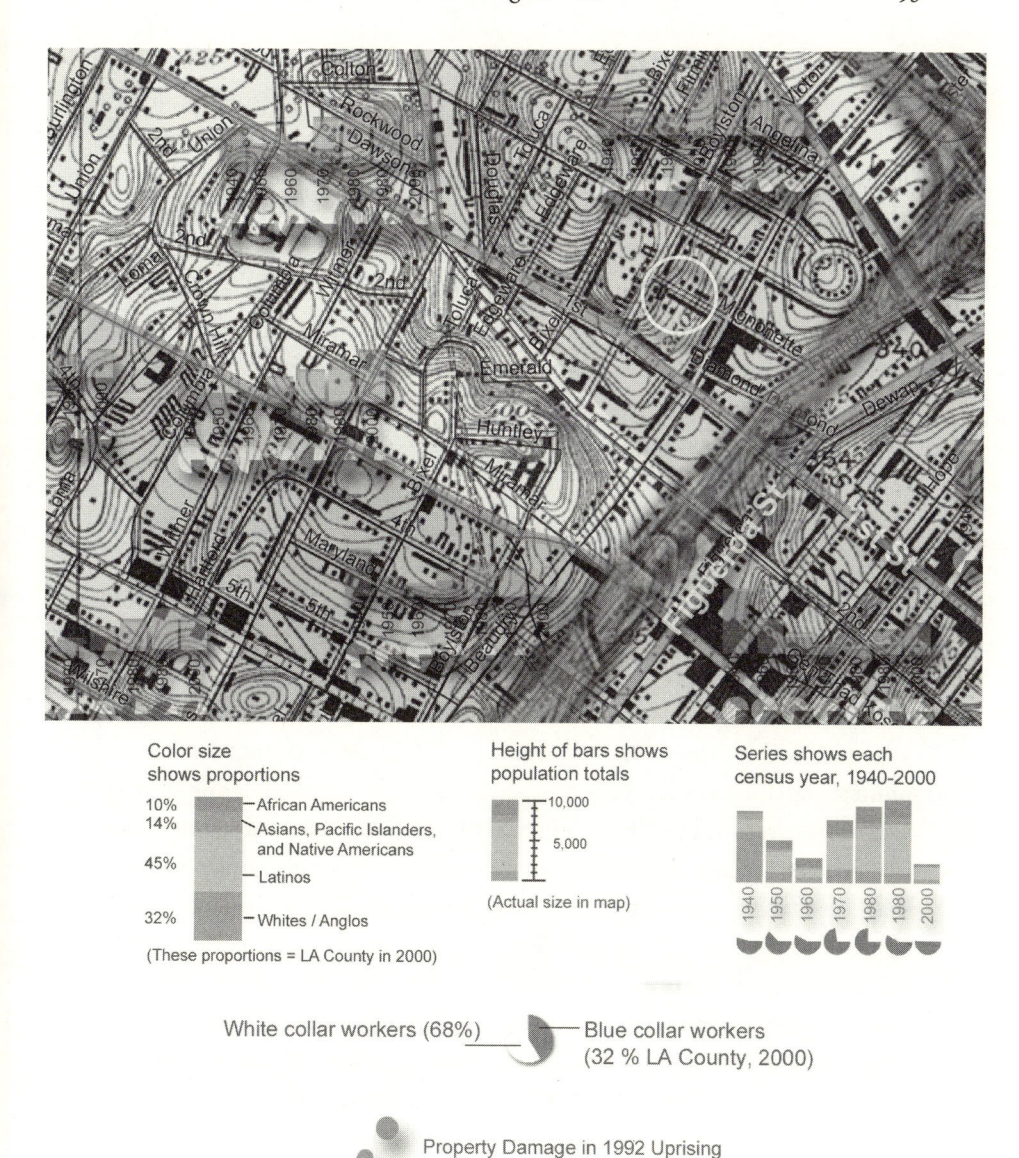

Inscribing the Past

Color size
shows proportions

10% — African Americans
14% — Asians, Pacific Islanders, and Native Americans
45% — Latinos
32% — Whites / Anglos

(These proportions = LA County in 2000)

Height of bars shows population totals

10,000
5,000

(Actual size in map)

Series shows each census year, 1940-2000

1940 1950 1960 1970 1980 1980 2000

White collar workers (68%) — Blue collar workers (32 % LA County, 2000)

Property Damage in 1992 Uprising

MAP 4.5. Detail of ghost map (Ethington, 2010), "Geveronga—Yaanga. El Pueblo de Nuestra Senor Reina de Los Angeles/La Placita/Downtown/Los Angeles, California, USA 0 CE–2000 CE" (Original in color). White circle indicates location of call-out, "Ghost Neighborhood: *Court and Beaudry, 1939 and 1997* (1997)" (figure 1). Cartography © 2010 Philip J. Ethington.

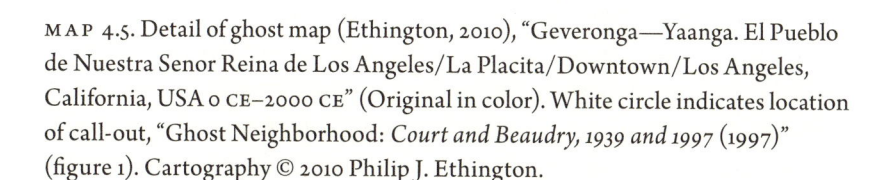

Today, historian and cartographer Ethington recounts anew the legendary deeds of the rulers of Los Angeles: dukes of petroleum, lords of world-destroying aircraft and aerospace technologies, and barons of motion pictures. Like Ekiken, he continues to inscribe the landscape with meaning. Unlike Ekiken, Ethington is not writing *fudoki*-style, purely textual narrative maps. He leverages the pictorial deep map to tell spatial narratives both pictorially and in prose. But, like Toyosawa interpreting Ekiken, who in turn interpreted the eighth-century "Records of Wind and Earth," others in the future will reference the inscribed landscapes of our present as their past.

CONCLUSIONS

Throughout this essay, we have explored the intertwining of historical depth with spatial narration. It should be clear from the Japanese and American examples that all places are deeply historical. The metaphor of "depth" speaks to the length of time that has passed since the earliest detectable human (or even nonhuman) actions inscribed into a particular landscape. *Inscription,* we have argued, is the attachment of meaning to a geophysical location. But inscriptions in this sense are potentially innumerable. They are not yet arranged into a story until someone creates a narrative from them. This is precisely what the Tokugawa scholar Kaibara Ekiken did. He assembled the stories of past inscriptions into a deep map—a textual map—in which he selected specific sites among many along the Saigoku Road between Kyoto and Arima. He also selected specific types of stories from each of those sites to recount in his 1711 *Arima no yama onsenki.* As the research of Toyosawa has revealed, Ekiken's textual map, despite its minimalist, seemingly factual, list-like format, holds a profound "hidden narrative," a story about Japan as a land of deities. Those deities are, in the tradition of Esoteric Buddhism founded by the eighth-century monk Kūkai, very ancient local gods who allegedly sought enlightenment by following Buddha. These local gods, in other words, were incorporated into an Indo-Chinese tradition that had been established as a unifying ideological regime for the Japanese archipelago. Ekiken had an ideological purpose in mapping the Kansai region the way he did: to promote the Japanese-centered history of places that had been influenced for centuries by Chinese culture. Ekiken's project was to reinscribe the deeds of *Japanese* deities and legendary mortals into these landscapes during the reign of the Manchus in China, when Japanese leaders were turning away from China as their model and consolidating the shogunate as its own center.

Toyosawa's explication of Ekiken's textual mapping is both textual and pictorial. She traces the textual references of the places mapped by Ekiken, back into the vast archive of Japanese literary traditions. She also inscribed Ekiken's sites into a cartographic representation of the landscape between Kyoto and Arima, to help understand the significance of the places he chose. We can say, then, that both Ekiken and Toyosawa mapped with a purpose. Ekiken's was protonationalistic, and Toyosawa's is critical—an attempt to understand Ekiken's role in inscribing Japan with ideologically laden narratives.

We have also shown, through the example of Ethington's ghost maps, that another strategy for deep mapping and spatial narrative is to construct pictorial maps that embed temporal change in visual vocabularies. Like Ekiken, Ethington seeks to make known the ancient past of a region that forms a vital part of a powerful state. Ghost maps make visible the ghostly traces of past regimes by representing the footprints of those regimes in the landscape. We have discussed the different kinds of representational power offered by the two different kinds of mapping: textual and pictorial. Ethington's ghost maps differ radically in form from Ekiken's *fudoki*-style writings. But they both have a very similar intent: to explicate what *took place*.

We have sought also to explain that textual and pictorial mapping are highly compatible. Each can do something that the other cannot. While Ethington's visual layering and sequential arrangement of historical data shows that narration can be achieved pictorially, even in a two-dimensional map, textual narratives are far more capable of conveying rich and extensive meanings. Any point within a pictorial map should be worthy of a lengthy textual narration of any of the many actions that took place there.

The net effect of these conclusions indicates that we are all readers of landscapes that are thickly inscribed with meanings. Understanding a landscape is to decode its deep narrative topography. It is to acknowledge that every act of mapping is to enter into a dialogue with the inscribed voices of the past, and to re-inscribe every mapped place for the future.

NOTES

1. Philip J. Ethington, "Placing the Past: 'Groundwork' for a Spatial Theory of History," *Rethinking History* 11, no. 4 (Dec. 2007), 463–530.
2. *Oxford English Dictionary Online* (accessed March 2005, July 2013).
3. Edward C. Tolman, "Cognitive Maps in Rats and Men," *Psychological Review* 55, no. 4 (1948), 189–208; K. Lynch, *The Image of the City* (Cambridge, Mass.: The MIT Press, 1960).
4. Henri Lefebvre, *The Production of Space*, trans. Donald Nicholson-Smith (Oxford, U.K.: Blackwell, 1991); Stuart Elden, "Politics, Philosophy, Geography: Henri Lefebvre in Recent Anglo-American Scholarship," *Antipode* 33, no. 5 (Nov. 2001), 809–825.
5. J. B. Harley, *The New Nature of Maps: Essays in the History of Cartography* (Baltimore, Md.: Johns Hopkins University Press, 2001).
6. Ekiken verifies this edict by quoting the excerpt from *Shoku Nihongi* (Chronicles of Japan Continued) in "Fusō kishō," in Kaibara Ekiken, *Ekiken zenshū* (Tokyo: Ekiken Zenshū Kankōbu, 1911), 7:310–542. Cited here at p. 311.
7. Akimoto asserts that the practice of calling regional gazetteers *fudoki* began sometime in the Nara period and was widely in use in the next era, the Heian (794–1185). Akimoto Kichirō, *Fudoki no kenkyū* (Kyoto: Mineruva Shobō, 1998), 42–44.
8. All *fudoki* from different provinces adhere to the instructions given in the 713 edict, but they differ from one another significantly and demonstrate each of the local characters. Generally speaking, they start describing the general features of the province, stating the size of the land, and number of shrines, followed by individual districts in the province. See, for example, Hashimoto Masayuki, *Kofudoki no kenkyū* (Osaka: Izumi Shoin, 2007) and Mark C. Funke, "Hitachi no kuni fudoki," *Monumenta Nipponica* 49, no. 1 (1994), 1–29.
9. *Fudoki* was introduced to Japan as part of centralizing efforts of the Yamato court that adopted much of Tang China's cultural, legal, and political systems. Yet, the court never converted Japan's political system or culture to the Chinese system entirely, and as the name *fudoki* reflects, the Chinese name for the regional histories was never adopted in Japan. Michiko Y. Aoki, *Izumo Fudoki* (Tokyo: Sophia University, 1971), 24–26.
10. Akimoto, *Fudoki no kenkyū*, 639. The eighth-century documents of the *Record of Ancient Matters* (*Kojiki*) and the *Chronicles of Japan* (*Nihon shoki*) record Emperor Ōjin as the fifteenth emperor who was a son of Empress Jingū, but his reign is controversial as the historical records overlap with another emperor.
11. In particular, *Izumo* and *Harima fudoki* have more descriptions and mentions of deities than other *fudoki*. For example, *Harima fudoki* explains the genesis of place names by (1) names resulting from some acts of deities, (2) names resulting people's acts toward deities, (3) names that originate in the presence of the deities (the deities were sitting there), and (4) names that originate in the deities that were sitting there. In this sense, *fudoki* captures the differences and characteristics of local provinces. Ibid., 847–848.
12. Japanese personal names in this chapter appear in the traditional order with the surname first, followed by the given name. The texts that Ekiken references in his spatial writings constitute the majority of the canonical works of *Kokushi taikei* (History of Japan, 1897–1904). Being a Confucian lecturer to the daimyo of Kuroda domain, while occasionally serving the shogunate in the shogun's capital of Edo, today's Tokyo, Ekiken

traveled frequently from his home domain to various places. These travels gave him the opportunities to access the latest books and scholarship.

13. With his decisive victory in the battle of Sekigahara in 1600, Tokugawa Ieyasu (1543–1616) became the first shogun in 1603 and established the Tokugawa shogunate that lasted until 1868. By 1650, new order had come to prevail, and the first half of the seventeenth century was when "the Tokugawa laid claim to a range of powers that made its government the strongest and most centralized in Japan's history." While the power of surveillance did not persist throughout the Tokugawa regime, Ekiken lived in a time when the shogunate's control over other daimyo was supreme. Constantine Vaporis, *Tour of Duty: Samurai, Military Service in Edo, and the Culture of Early Modern Japan* (Honolulu: University of Hawai'i Press, 2008), 13.

14. In this essay, all translations from Japanese are by Nobuko Toyosawa. Kaibara, "Fusō kishō," 311.

15. These contemporary scholars include Matsushita Kenrin (1637–1703), Kurokawa Dōyū (d. 1691), who wrote *Yōshū fushi* (Record of Yamashiro Province, 1684), and *Yamashiro kokushi* (Record of Yamashiro Province) by Ōshima Kyūma. Ibid., 333.

16. As Toby points out, for Ekiken to write Japanese history, he had to first confront Chinese discourse on Japan, such as the *Houhanshu* (History of the Later Han) that dates as early as the year 57 CE, which mentions Japan. However, the Chinese discourse on Japan included misguided and erroneous views on Japan because they were written from the perspective of the outsider. Ronald P. Toby, "Foreign Texts/Native Reading: Matsushita Kenrin (1637–1703) and the Challenge of Chinese/Korean Histories," in Martin Collcutt, Mikio Katō and Ronald P. Toby, eds., *Japan and Its Worlds: Marius B. Jansen and the Internationalization of Japanese Studies* (Tokyo: I-House Press, 2007), 143–174. Cited here at p. 148. See, also, Kate W. Nakai, "Tokugawa Confucian Historiography: the Hayashi, Early Mito School and Arai Hakuseki," in Peter Nosco, ed., *Confucianism and Tokugawa Culture* (Honolulu: University of Hawai'i Press, 1997), 62–91; Kate W. Nakai, "The Naturalization of Confucianism in Tokugawa Japan: The Problem of Sinocentrism," *Harvard Journal of Asiatic Studies* 40, no. 1 (1980), 157–199.

17. We draw the idea of the reading public from Yokota Fuyuhiko who examines growing readerships and students who studied *Ekiken-bon* (Ekiken books) among the wealthy farmers in the village in the eighteenth century. Yokota Fuyuhiko, "Ekiken-bon no dokusha," in Yokoyama Toshio, ed., *Kaibara Ekiken: tenchi waraku no bunmeigaku* (Tokyo: Heibonsha, 1995), 315–353. See, also, the following works about the reading public in Tokugawa Japan by Yokota Fuyuhiko: "Kinsei minshū shakai ni okeru chiteki dokusha no seiritsu: Ekiken bon o yomu jidai," *Edo no shisō: dokusho no shakishi*, no. 5 (December 1996), 48–68; "'Rōnin hyakushō' Yoda Chōan no dokusho," *Hitotsubashi ronsō* 134, no. 4 (2005), 612–635; "Edo jidai minshū no dokusho," *Rekishi chiri kyōiku*, no. 718 (August 2007), 76–81.

18. Kyoto Shoshi Hensen shi Hensan Iinkai, ed., *Shuppan bunka no genryū: Kyoto shoshi hensen shi, 1600–1945* (Kyoto: Kaiseisha, 1994) and Munemasa Isoo, *Kinsei Kyoto shuppan bunka no kenkyū* (Kyoto: Dōhōsha Shuppan, 1982).

19. One *ri* is about 3.9 kilometers, which is about 2.4 miles.

20. Kaibara Ekiken, "Arima no yama onsenki," in Kaibara Ekiken, *Ekiken zenshū*, 7:268–292. Cited here at p. 271.

21. Kūkai was the first to hold a position that a human is originally enlightened and one can attain enlightenment here in this life. He fortunately obtained patronage from

Emperor Saga (786–842, r. 809–823), who entrusted the completion of the Tōji Temple to Kūkai, allowing him to make the temple the center of the Esoteric Buddhism in Kyoto. Kūkai's entry into the Tōji Temple meant the establishment of his religion, the Shingon sect of Buddhism, the term approved and given by Emperor Jun'na (786–840, r. 823–833). Kūkai's religion had many elements compatible with the indigenous deity worship, in particular, man's oneness with nature, and developed very closely with the mountain asceticism. Yoshio S. Hakeda, *Kūkai: Major Works* (New York: Columbia University Press, 1972), 6–7, 54–55.

22. Kaibara, "Arima no yama onsenki," 273.

23. Ibid., 277.

24. According to a local legend, the deity *Atago gongen* disliked those who came to Arima, accompanying military bows called *shigetō yumi*, white arrows, dappled gray horses, or hawks, and resulted in bestowing misfortune upon them. Ibid., 281–282.

25. The practice of worshipping these deities at Buddhist temples spread widely among popular believers in the medieval era. Emerging from the fusion of foreign religion Buddhism and indigenous belief of worshipping native deities, *kami*, this practice identified the *kami* with the Buddhas and Bodhisattvas on the theory that the *kami* were incarnations of the Buddhas and Bodhisattvas. This doctrine, known as *honji suijaku*, and other forms of fusion of Buddhism and Shinto, were largely made possible by the establishment of Esoteric Buddhism as a system of thought. Hakeda, *Kūkai*, 8. For a more detailed analysis on the relationship between medieval ideology and practice and the system created during the Heian era, see Allan G. Grapard, "Institution, Ritual, and Ideology: the Twenty-Two Shrine-Temple Multiplexes of Heian Japan," *History of Religions* 27, no. 3 (1988), 246–269.

26. While Ekiken does not provide the reader with detailed explanations or the history of each place, to proclaim that *Yu no gongen, Miwa myōjin*, and *Kashita myōjin* are three tutelary shrines for the Arima Mountain implies that he has investigated the history of shrines by consulting such documents as volumes 9 and 10 of a fifty-volume work of *Engishiki* (Regulations and Laws of the Engi Era, 927), which list registered shrines in the imperial palace, Kyoto, the five home provinces, and seven circuits. In total, there are 2,861 names of shrines listed in these volumes, indicating that they existed and were registered with the central government before the compilation of the document. Ekiken only includes minimum information for the purpose of this guidebook, and his topographic and historical investigations are found elsewhere. Kaibara, "Arima no yama onsenki," 282.

27. The pilgrimage route dates from the eleventh century, and this is one of the oldest and most popular pilgrimage routes known in Japan. See, for example, Maria Rodríguez del Alisal, Peter Ackermann, and Dolores Martínez, eds., *Pilgrimages and Spiritual Quest in Japan* (New York: Routledge, 2007); Tōbu Bijutsukan, Kyoto Bunka Hakubutsukan, and Nihon Keizai Shinbunsha, eds., *Saigoku sanjūsansho: Kannon rejiō no shinkō to bijutsu* (Tokyo: Nihon Keizai Shinbunsha, 1995).

28. The explanatory notes to volume 3 of *Ekiken zenshū* state that *Jingikun* is a text published sometime prior to the Kyōhō era (1716–1736) by Publisher Ryūshiken in Kyoto, but the surviving version today is one of the variant texts. However, the editors examined several variant texts to make the copy included in the volume to the state of "relatively complete." Ekiken Zenshū Kankōkai, ed., "Hanrei," in *Ekiken zenshū* (Tokyo: Ekiken Zenshū Kankōkai, 1911), 3:1–3. Cited here at pp. 2–3.

29. Kaibara Ekiken, "Jingikun," in E. Kaibara, *Ekiken zenshū* (Tokyo: Ekiken Zenshū Kankōkai, 1911), 3:641–685. Cited here at p. 668.

30. Ibid., 669.

31. Denis Wood, *The Power of Maps* (New York: Guilford Press, 1992). Wood elaborates the perspectival, historical, and selective qualities of pictorial maps; John L. Gaddis, *The Landscape of History: How Historians Map the Past* (New York: Oxford, 2002) makes many similar points about cartographic mapping as an act of reduction, abstraction, scale-jumping, and narration, and shows that historians writing textually narratively carry out identical or very analogous acts of concision.

32. *Ghost Metropolis* is a global history of Los Angeles since earliest human habitation, presented as a hybrid of textual, cartographic, and photographic representation, in print, online (HyperCities), and public art formats. *Ghost Metropolis* is a twenty-first-century "atlas," inspired by the Renaissance atlases of the sixteenth and seventeenth centuries: rich mixtures of typography, graphic arts, and cartography. Ghost maps are one part of Ethington's efforts to expand the range of the historian's craft, by recounting the past through cartography and visual media *alongside,* and *in combination with,* textual narrative.

33. While we know that the region was inhabited continuously for ten thousand years by a single culture, the so-called Millingstone People, we do not know what names they gave to themselves nor to their inhabited places. Thus, Geveronga and Yaanga are the first known place names for this particular "downtown" area, given by the Uto-Aztecans, who arrived approximately at the time of the Western Han Dynasty or the Roman Civil War: approximately two thousand years ago.

34. Author's revision of the Loeb Classical Library English translation of Ovid's *Fasti*, by Sir James George Frazer, revised by G. P. Goode, second edition, 6 volumes (Cambridge, Mass.: Harvard University Press, 1976). Quotation from Latin on p. 266 and English on p. 267, lines 93–94.

35. "The *United States Census Bureau* has designated the five county region as the Los Angeles-Long Beach-Riverside, CA combined statistical area, with a January 1, 2009 population estimate of 17,786,419." Quoted from "Greater Los Angeles Area," http://en.wikipedia.org/wiki/Greater_Los_Angeles_Area (accessed April 30, 2012), which cites: "Annual Estimates of the Population of Combined Statistical Areas: April 1, 2000 to July 1, 2009" (CSV). *2009 Population Estimates* (United States Census Bureau, Population Division. March 23, 2010). Retrieved March 29, 2010.

36. This paragraph and the following four paragraphs are drawn from Philip J. Ethington, *Ab urbe condita:* The Regional Regimes of Los Angeles since 13,000 before Present," in William Deverell and Greg Hise, eds., *A Companion to Los Angeles,* Blackwell Companions to American History (Cambridge, Mass.: Blackwell, 2010), 177–215.

5

QUELLING IMPERIOUS URGES

Deep Emotional Mappings and the Ethnopoetics of Space

STUART C. AITKEN

INTRODUCTION

Questions about the setting in virtual motion of spaces, times, and technologies by and through bodies form a core discussion in geography. The arts and humanities have been inextricably linked to understanding these events since at least the mid-seventeenth century, but the last century in particular witnessed a mighty series of spatial and technological turns. In 1935, Walter Benjamin responded to these turns by proposing the marriage of space, place, and time as a product of modern technological reproduction. He argued that reproduction of this kind "orients knowledge towards the public sphere, but also simultaneously orients the public sphere towards that knowledge."[1] When he talks of knowledge, Benjamin means something *material* and *lively*, not abstract and objectified. In short, there is an emotive and visceral connection among technologies, spaces, and time, which for Benjamin is practiced in the public sphere. A little later but in a similar vein, psychologists and critical culture theorists such as Jacques Lacan, Henri Lefebvre, Guy Debord, and Fredric Jameson argued that places operate as social and psychic mechanisms, creating their individual and social subjects out of space and time. The idea of identity creation divorced from space, place, and technology was at once untenable and unthinkable. It seems to me that this connection among identity, space, place, and technology is at the heart of those parts of the Marxist Situationist International movement that fed into the May

1968 riots in France and spawned, among other things, the term *deep mapping*. As David Bodenhamer points out, deep mapping is about spatial stories, memory, images, knowledge, and identity.[2] He quotes Pearson and Shanks who argue that deep mapping "attempts to record and represent the grain and patina of place through juxtapositions and interpretations of the historical and the contemporary, the political and the poetic, the discursive and the sensual. . . ."[3] In short, it is a push to map moments, movements, and pieces of humanity through emotions, poetics, and the political.

Through deep mapping, places are unpacked in terms of the modern subject's spatial and political unconscious as well as layers of memories.[4] And there is an important reciprocation between the liveliness of Benjamin's technological transformations and the reality of day-to-day living. Using more and more sophisticated technologies, deep maps transform yesterday's two-dimensional maps and simulated virtual representations into today's real virtual environments that are simultaneously multidimensional, multivalent, and sensory. Today the visceral and the material are very much apparent and connected publically, and this connection is spatial and experiential. Today's deep maps arise out of virtual materials that project multivalent environments that have capacity *en abyme* (literally, into the abyss).[5] But, as Giuliana Bruno, Elizabeth Grosz, and others have pointed out, the connections between technology and art, and places and power are by no means straightforward and progressive.[6] Grosz argues that identity is created through processes that are virtual (from Deleuze she points out that our identity is always what we do in the virtual ever-present) and through a variety of forces and valences, the most fundamental of which relate to what she calls *geopower*. Put simply, the earth is framed through geopower, which creates a condition for the plane of composition and thus for particular works of art. Framings cut through territories, break up systems of enclosure and performance, traverse territories, and then reconnect with chaos, enabling something of the chaos outside to reassert and restore itself in and through bodies and their works. Deep maps are a form of framing that, to paraphrase Grosz, create and metabolize sensations that are released into the world and made to live a life of their own, to infect and transform other sensations.[7] Bruno argues that a large part of the energy that elaborates these processes comes from the emotive contexts of bodies and movement that, as Benjamin pointed out years ago, creates a quizzical public. And Benjamin's arguments join Bruno's recognition that the "public," oftentimes "has neither yardsticks for its judgments nor a language for its feelings."[8] With this essay, I want to explore the visceral, virtual, and embodied through a mapping of feelings that may move us some way into geopower and beyond to perhaps create Benjamin's *language for feelings*.

My concerns begin with a push against the creation of imperious subjects through cartographic design that is two-dimensional and intent upon disempowerment. The link among geography, mapping, and empire is well established, as is the connection between placated subjects and memory in what Michael Rogin terms the "spectacle of amnesia in imperial politics." Rogin's thinking mirrors Guy Debord's notion of spectacle as a fake reality that hides capitalism's degradation of human life.[9] To the extent that authentic social life is replaced by representation (Benjamin's mass media), so cartographic images are part of social relations that are mediated by spectacle. Pushing through Apollonian desires for control and the subjugation of memory, it seems to me that cartography also offers the possibility of Dionysian redemption through Bruno's *carte du pays de tendre,* or tender mappings. There is a fundamental geopower to these mappings that give voice to a language for feelings and memory.

With this essay, I want to create a cartographic and poetic language for feelings and memory that make some sense of the complexity, panic, and hysteria induced by environments *en abyme,* that is, at the edge of the imperial abyss. I do so with what I call *ethnopoetry,* which is an emotive mapping of stories that describe connections of people to other people and to places. I do not claim the annihilation of the altogether overwhelming imperial abyss (nor would I want to because it is out of this chaos that creativity and art arises), but the possibility of redemption—at least for a moment—from our imperious subjectivities. In what follows, I lay out a navigational strategy through contemporary lives and landscapes, drawing inspiration from Grosz's geopower and Bruno's reworking of imperial cartographies in the form of journeys through tenderness, and her re-visioning of imperial subjects into voyagers streetwalking on a ruined map.[10] This chapter ends with an ethnographic example of a father's journey, which helps tie together some of the outlined theoretical threads.

TECHNOLOGY AND REPRESENTATIONAL CONTROL

To this day, the British Ordinance Survey (OS) produces some of world's most beautiful maps: carefully symbolized, closely colored contours edge nicely hachured mountains to produce an engaging representation of place. As a teenager living in Scotland, these maps drew me in, stimulating my geographic imagination. What I did not know at that time was that, as a precursor to the first OS production, the *Duke of Cumberland's Map* (1747) lived in infamy as the technology that facilitated the subjugation and genocide of the Scottish people after the 1745 rebellion and the defeat of the Jacobites at Culloden. The Duke of Cumberland was hated by many Scots for his map and the Highland clearances that occurred in the wake of its creation. The hysteria of that time was, it can be argued, derived in large part by the use a technology of representational control. Cumberland's infamous map and the military survey that followed suggested the possibility of precise identification of where the Jacobites lived, how they could be extracted, and what resources could be put into the service of empire. The map was at most a powerful representational suggestion as there is no evidence that it was actually used to root out Jacobites or facilitate the Highland clearances.[11] It was nonetheless popularized as a way of controlling a rugged and unreadable landscape populated by wild and dangerous people.

Rachel Hewitt starts her celebrated book on the OS with the story of how Charles Edward Stuart, in hysterical grief after the failure of his rebellion, confided to the elder Lord Lovat that he was undone and wanted nothing more than to escape the country. The two parted company and Lovat made his way alone to Loch a'Mhuillidh. When cornered by English Redcoats, the old, tired, and infirm Lovat disappeared southwest into the seeming impenetrable clusters of peaks and Munros that he had known since a boy. The Redcoats were beaten back by the formidable and unreadable landscape. As other Highland lords frustrated capture in what appeared to the English as impenetrable terrain, Cumberland realized that his task was exacerbated by inadequate maps of the region. The map that came to bear his name was created to control and master the Highland terrain, and also was created to stave off anxiety over Highlander guerilla tactics that went back to the wars of independence in the thirteenth century.[12]

I argue elsewhere that desire and anxiety are bound in a psychological struggle to maintain balance between "manhood" and masculine subjectivities on the one hand and, on the other hand, power and dominance over the projected and represented other.[13] The key forces of desire and anxiety are *mise-en-abyme* (literally "placed into the abyss") within which male power and its imperial urges fitfully repose. But the construction of the imperial abyss is also about the disruption of the defenses set up to protect the boundaries between spaces of the self and spaces of the other, and so it parodies desire and anxiety for the oceanic oneness of the preoedipal/premirror stage where there is purportedly no separation between subject and object. For Grosz, this is where geopower resides and it is a force that is precisely and always about the production of art, but the elegance and beauty of the artistic is usurped by patriarchy and empire, and those processes are well delineated by Freud and, later, Lacan. The Freudian/Lacanian context of sexual desire, repression, and empire takes geopower in a different direction from the artistic and emotional lines-of-flight proposed by Grosz and Bruno. For Grosz and Bruno, it is from placement within the imperial abyss that hope arises. With the *mise-en-abyme* there is a collapse of the scalar fix in subjectivity; the subject is not only in crisis (in accordance with the norms of hegemonic masculinity), but is fluid and able to float outward and inward without resistance. This is not a moral issue, but rather a permutation of virtual identity, beauty, and art. Not only is the hegemonic subject in crisis but the reality *produced* by the subject is also in crisis. Hysteria at the edge of the abyss simultaneously undermines the hierarchies of patriarchal space, penetrates the fixity of patriarchal scales and imperial spaces, and transgresses traditional social and identity politics with the virtual. Along these lines, Henri Lefebvre points out that the state in the twentieth century crushes time "by reducing differences to repetitions or circularities . . . it neutralizes whatever resists it by castration or crushing . . . state-imposed normality makes permanent transgression inevitable."[14] From these sites, viewers and characters can interpret, and contest, dominant narratives and mappings as they are woven into obsessively repetitive social patterns and naturalized interpersonal relations.

Technology enables transgression because it can provide a large swathe of people access to spatial data and access to analytic tools at reasonable prices. This is the wonder of any technology: it has a hugely democratizing potential that can open and rekindle public awareness, participation, and action. In this sense, it works against Rogin's spectacle of amnesia and the creation of imperious subjects. Witness the speed with which the Reformation spread through sixteenth-century Europe, using Gutenberg's technology. The invention of the commercial printing press in Mainz, Germany, revolutionized how knowledge and information were shared. It did not take long for Erasmus's *Novum Instrumentum* (1516) to demonstrate the corruption of the Latin Vulgate's New Testament. That new technology—the *instrumentum*—enabled translation of the Bible into "vernacular tongues" for reading by "the ploughboy" and "the simplest woman."[15] Witness two Mexican uprisings—one in the state of Morelos that culminated in the revolution of 1910, and one in the state of Chiapas in 1994. In the former, Emiliano Zapata used the new technology of railroads to move his guerilla fighters and quickly take control of the northern Mexican states. In Chiapas, the 1994 inauguration of North American Free Trade Agreement inspired contemporary Zapatistas to challenge the image of Mexico as a modern, youthful nation, eager for change and unencumbered by centuries of rural poverty and oppression. The technology they used in this challenge was the internet.[16] Am I suggesting that deep mapping has the propensity for revolution akin to the European Reformation and Mexican revolutions? History, for what it is worth, will judge better than I, but I nonetheless believe that these technologies are fomenting important spatial—and hence social and political—changes to which we need to heed. Space—its representations and affects—is opening up in a public flurry of democratization. Using geographic information systems and global positioning systems, lay people can visually construct, deconstruct, deterritorialize, and reterritorialize spatial realities, and the result is a profound experiential and epistemological shift. The issues raised are profoundly political: there is no longer an unproblematic and empirically verifiable "space" to which we can refer. Referentiality and legitimation without accountability to larger democratic changes are no longer tenable. The weakened representational function of space (of maps, of visual images) produces little sense of permanence, history, or material investment. What is left is aesthetic, affective, and openly political. This is, without doubt, a good thing. It is, of course, not a new thing.

MOVEMENTS, EMOTIONS, AND MAPPINGS

Weakening the representational function of space is always good. Cartographic representations in stasis are dangerous. As framings, static maps work to foreclose the possibility of alternate interpretations via an emphatic rendering of "naturalness" or "inevitability"; they may even, as Rogin argues, attempt to forestall the cognitive act of interpretation itself.[17] But they are also profoundly destabilizing as suggested by the representational psychology of the *mise-en-abyme* described earlier. The production of identity is a matter of the co-construction of psychological and state apparatuses which, for Bruno, emerges through the processes of intertextuality and slippage. Bruno's voyager is by no means a passive, malleable creature. Ultimately, an emphasis on slippage prompts certain skepticism in regard to surety, stability, rootedness, and order. Bruno engages the contexts of mapping in architecture, travel, design, housing, planning, and film. More pointedly, she takes the history of mapping and contextualizes it in the arts, in desire,[18] and in tenderness.

Taking the *Femme-Maison* (1947) drawings of French artist Louise Bourgeois as an example, Bruno notes how the house and body are joined.[19] The house/woman has legs and looks like she is waving as if, perhaps, departing on a journey. The work challenges long-standing notions of home and gender fixity. The home is no longer enclosed or enclosing. The drawings:

> show the motion of emotions that reside in-house. In this affective mapping drawn by women, home can indeed turn into a voyage.... The vision of *emotion* inspires us to consider the relation of voyage and dwelling, in order to see how sexual difference can be housed differently. To look architecturally, and with geographical eyes, at the relation of private to public space can advance former notions of gender identity based on psychoanalytically oriented feminist theory... as a geography of negotiated terrains. Thinking geographically, we can design different cultural maps as we venture into ... a lived space.[20]

Bruno calls Bourgeois's reworking of cartographic themes against prevailing patriarchal and imperial hegemonies a "sentimental geography."[21] Bourgeois's recent work continues her interest in the deterritorialization

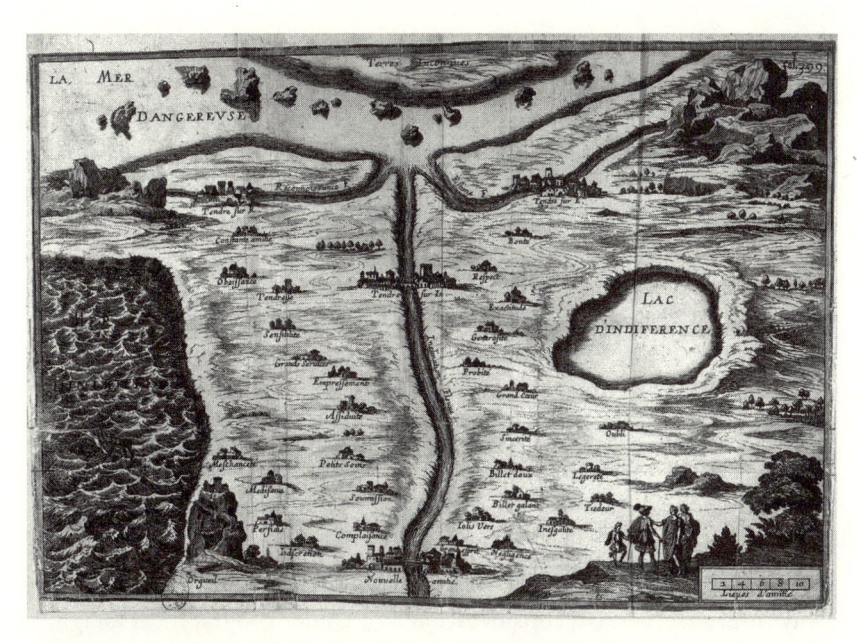

FIGURE 5.1. Madelaine de Scudéry's *Carte du pays de Tendre.*
Engraved by François Chauveau (1654).

of gendered contexts. Her *Arch of Hysteria* (1993), for example, provides a clear-cut and evocative reworking of sexualized boundaries that speaks to the very source of the twentieth century's plays on hysteria as a manmade woman's disease.[22]

In her *Atlas of Emotion,* Bruno evokes Madelaine de Scudéry's map that accompanies the 1654 novel *Clélie.* Scudéry's *Carte du Pays de Tendre* is a celebrated allegory for the female association of desire with space, and an exemplar of the ways in which cartography is inextricably linked with the shaping of a female subjectivity (figure 5.1). It highlights important passages and mobilities away from lakes of indifference, dangerous seas, and *terra incognitae* to favorable villages and towns of tenderness, large hearts, reflection, and sympathy.[23] Tom Conley suggests that the map in *Clélie* might have been drawn in opposition to contemporaneous military cartographies and technologies, inaugurated by neo-Cartesian engineers under kings Henry IV through Louis XIV who redrew the defensive lines of France and designed fortified cities at a time when new siege technologies were changing the ways of waging war.[24] The map shows a geography entirely based on emotive themes of love, fear, indifference, and so forth. Following the themes of love, for example, a River of Inclination flows past the villages of *Billet Doux* (love letters) and *Petits Soins* (little trinkets). The way through this pastoral country of affections begins at *Nouvelle Amitié* and leads (ignoring dead ends such as the Lake of Indifference) by three alternative routes to either *tendre-sur-reconnaissance, tendre-sur-inclination,* or *tendre-sur-estime.*[25] The map was very much part of the antiwar movement that grew up at the time in the salons of Paris. It became a popular symbol for the politically and culturally independent, aristocratic *salonnières.*[26]

Bruno's arguments shift the context of visual geographies and the art and science of war; her purview includes and goes well beyond maps, paintings, and films; it moves away from patriarchy and imperialism—away from surety—to a consideration of a tender geographical imagination. In this extensive reworking of cartography away from its imperial roots, Bruno offers a reterritorialization based on motion and emotion. Scudéry's map graphically embodies the central tenet of Bruno's argument that is embodied in the traveler: that motion produces emotion and that emotion contains movement. Although she never defines what she means by emotion, Bruno's tactile, tender mapping is nonetheless an allegory for female association of desire with spaces that are not referential to cognitive or psychoanalytic theory. It moves discussion away from the binary oppositions of home and travel, sedentarism and movement, and the problematic connections between home and well-being, and the nurturing woman in her place. In making this point, Maria Walsh sees Bruno concentrating on a "mapping where exterior and interior collapse on the terrain of a singular topology."[27] Walsh goes on to point out that Bruno's reluctance to define emotion does not detract from an argument that presupposes an important intertextuality between the museum, the art gallery, the building, the movie theater, and its phantasmagoric precursors that relates to memories that affect and connect us, rather than divorce us from real connection as sites of alienating spectacle. Herein lies a remedy to the illness of the imperious subject that is tied to movement, desire, poetics, and politics.

THE ETHNOPOETICS OF SPACE

By spending time with the emotional and affective relations among space, art, and architecture, Bruno opens the possibility for a larger discussion of technology, art, and space. Spaces are lyrical events to the extent that they emplace and transform our connections to other people and the world around us in ways that are perhaps more centered, perhaps more spiritual, perhaps more politically charged than our run-of-the-mill daily grind. Landscape artists, musicians, novelists, and poets often tap this energy, which is precisely what Grosz means by geopower. If we think of spaces as events and events encourage change, then spaces and people *become* through complex, geopowerful relations. In the example that follows, I describe these relations lyrically and poetically so as not to diminish the importance of emotion as a political push in the world. In larger debates in the sciences and humanities to which Bruno, Grosz, and other feminists contribute, it is clear that emotions matter. They affect the way we sense the past, present, and future, and they help us negotiate events. If poetry is an emotive construction of language, then my arrogance is to revisualize, contort, and arrange people's words and gestures to create something that speaks to their emotions. I call this *ethnopoetry,* because it is derived from discussions, stories, narratives, biographies, interview transcripts, participant observation, and experimentation with visual and playful methodologies.[28]

I want to suggest a way of negotiating spatial politics through experiencing material relations as ongoing, affective, and embodied. Spaces pose in particular form the question of how biographies and power are negotiated. Understanding the embeddedness of power relations in space helps us to see how change plays out in particular locales, but considering power alone is insufficient to understand the dynamic relations between people and places because, as a habituated aspect of our lives, spaces are often envisaged as contexts through which events play out. This tame and passive notion of space belies the affective and transformative properties of geopower. An active notion of space positions it as permeable and fluid: space as an event. If we consider spaces as events in the sense that they are an assemblage of previously unrelated forces rather than a thing, then it changes the way we think about spatial politics.

The ethnopoetics of space contrive a representational method that focuses on space as an event rather than a static stage. In the example that follows, I use images, dialogue, narrative, and poetry to provide a parsimonious rendering of the emotions that exceed the text. I work at the words, pushing them to reveal the emotional power of a conversation, a meeting, a visual rendering. I want to get at the embodied power that resides in people and places. It is about the language of looks, twitches, grimaces, tears, laughter. It is about connections and places and past remembered conversations. Ethnopoetics are a halting and partial—and yet sincere and promising—attempt to represent the nonrepresentable of a conversation, a narrative, a life. As a lyrical experiment, I use different genres—mixing poetry with dialogue, images, and academic discussion—to transform stories into deep mappings. Through the silences that join and link the narrative and the poetry, it is my hope that emotions foment and reveal themselves. These are the collective experiences of change, the content of which is always partial and incomplete and always generative and creative.

The term *ethnopoetry* was coined by poet Jason Rothenberg in collaboration with anthropologist George Quasha in a 1968 volume titled *Technicians of the Sacred*.[29] Six years after the publication of *Technicians of the Sacred*, Rothenberg (with Michel Benamou) convened an international symposium on ethnopoetry in Milwaukee that was, in part, a foundation for a new journal entitled *Alcheringa: Ethnopoetics*. The journal got off to a shaky start but went on to publish nine volumes of ethnic poetry between 1970 and 1980. *Alcheringa* is aboriginal for "dreamtime," and many used the new field as a basis for taking oral traditions and attempting to stay cultural genocide with codifications and classifications that use Western traditions. Others eschewed Western traditions of linearity and objectivity, if not writing, to elaborate the emotional and circular performance of oral traditions. What is more telling is the way Rothenberg and Quasha report at the symposium on the potential of ethnopoetry.

Rothenberg notes that *ethnos* was not always taken as groups in isolation, centering, orbiting around some Western standard but was, rather, an expression of otherness: "a sign that points from what we are or may become to what we aren't, haven't thought ourselves to be, may fear or scorn."[30] At that earlier time, he argues, *ethnos* meant nation, people, group, or race, not as *this nation* but as *those nations*. *Ethnopoetics* by this definition is not a new construction but a reminder of an older truth or linkage and, importantly, as poets *we* are *them*. Rothenberg's definition of *ethnopoetics* moves against the classificatory linguistics that attempted to capture the term.[31] Poetics, Rothenberg goes on to argue, is the clincher for something different. It is a process of creation, of "coming into knowing where we are. To say, articulate, our sense of being in the world, however changeful, dangerous and slippery ... [ethnopoetics] struggles to make, create, an instrument of language, discourse, are, to map the changes, to facilitate them, live in the hope of transformation, of a deepened, heightened, sense of who we are and where." Rothenberg was talking about mapping and transforming ourselves and other cultures, or through other cultures, by documenting in some way their oral histories. He cautions that by articulating academically conversations through a poetry that speaks to culture, much is destroyed in an attempt to build a bridge.[32] As the crisis of representation theorists of the 1980s pointed out, the mapping is about us rather than the other cultures. What I try to do differently here is cocreate the art of the event of space. What follows is an example of a story about fathering that I cocreated with a participant in a fifteen-year ethnographic project.

FATHERING FINDING FORM: AN EXAMPLE OF TENDER MAPPING

The coproduction of bodies, memories, emotions, and space matter. For a number of years now, I have been asking questions regarding men's resistance to and culpability with hegemonic norms of fatherhood.[33] How, precisely, do men struggle against formative notions of an imperious and patriarchal norm that prescribes the "law of the father"? What spaces facilitate this? The example I have chosen to elaborate an ethnopoetics of space, comes from an ethnographic project with fathers that spans fifteen years. During that time, I got to know Rex and his story as a father. When I met him, he was in his late forties with two estranged sons, Manny and Simon. For a large part of his life, Rex was a "line man" for San Diego Gas and Electric, working on telephone poles and power lines. After successfully battling substance abuse, Rex earned a master's degree in social work. One of Rex's passions was reading and he particularly enjoyed debate on issues of social and labor justice. Over the years I knew him, his relations with his sons improved and, as they did so, his passions turned to his two boys and his grandson. Appendix A is an elaboration of a transcribed interview that I conducted with Rex in 2008, two years before he died.[34]

The boxed dialogue and discussion (Appendix A) is part of a larger three-hour interview in which Rex took time to think through his fathering story. The story is partitioned chronologically through a series of moves from one house to another, from a beach community to a mountainous interior community and then back to a neighborhood close to San Diego State University: seven moves in all. To this extent, Rex's story is about movement, relocation, and the search for a place for his fathering that is not an imperious connection with his boys. That said, there is still the continued need for connection through control. The rural idyll does not work because, for Rex, that geographic solution is not one about which he has control. His continued moves to find a better place to raise Manny after Simon and his first wife Molly left certainly reflect his commitment as a father, but it is the emotional trauma brought on by his son's threat of leaving that refracts the loss of Simon years before and highlights for Rex what is truly important. And it is the beginning of his move away from trying to control the people around him. The pain of the possibility of losing Manny to his continued abuse of alcohol was too much for Rex, and so he was pushed to find a solution that was not about geographic change but inner change. This was the crux of fathering for Rex and the emotional precipice (perhaps a *mise-en-abyme*) upon which his whole sense of self teetered. Figure 5.2 reflects Rex's physical movement through Cartesian space by unlabeled arrows. The other arrows represent the departure of Molly to Ramona, Simon to Nebraska, and Manny to Los Angeles.

Rex's boys moved on and created families of their own. Manny is married with no kids and Simon is divorced and raising a son. Toward the end of his life, Rex created an important material space to the connection that he enjoy with his boys. It brings together his belief in what it means for him to be a father, which follows at least in part the material practice of teaching. I ask Rex about his connection with Simon and his grandson and get an inclination of a connection that is primarily an emotional bond but is reified in Rex's idea of fatherhood. It is a fathering form that is ma-

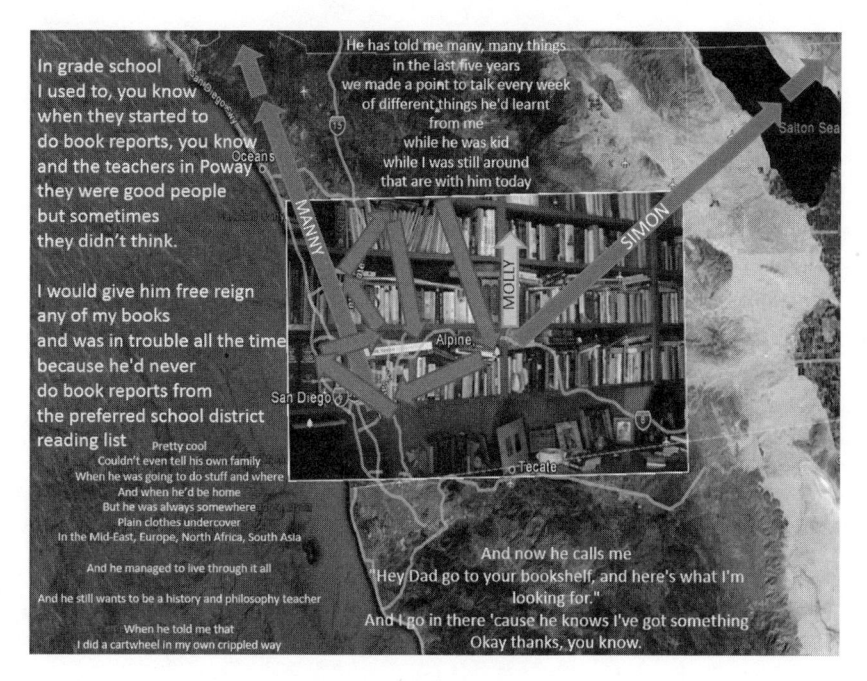

FIGURE 5.2. Ethnopoetics 2: Fathering Finding Form.

terial, sedentary, and finds form in a bookcase. Pulling from Bourgeois's *Femme-Maison,* figure 5.2 describes a moving journey that circumgyrates through and beyond Rex and his books. These are books that enabled Rex to support a left-leaning Simon through a notoriously right-wing school district in Poway: poem at top right of figure 5.2. These texts stand out as poetic cartouches that touch on the work of fathering. They suggest a father's sense of accomplishment after an emotionally trying journey. Bottom right: Manny would call his dad from Los Angeles and ask him to look something up. The crowning accomplishment for Rex comes with the text at the bottom left. After spending a number of years with Special Forces traveling around the world, Simon returns to look after his son in Nebraska. Simon, who left in anger and did not talk to Rex for eight and a half years, began talking to his dad on a weekly basis. Simon wants to be a history and philosophy teacher. The poem in the lower left quadrant of figure 5.2 speaks to a visceral joy that takes over a man whose body is semicrippled by years of work on telephone poles.

Rex created a messy and convoluted family community that moves through a series of phases involving his sons, his parents, his ex-wives, and, latterly, his grandson. It comes together and separates at various points, it involves movement across, and the creation of, space; it involves surprises and the happenstance along with the planned and some elements of the predictable. Rex was the first to admit that for the most part serendipity and thrown-togetherness win out over planning and predictability. What happened for Rex was also a series of repetitions that did not address the root of his fathering problems. The failure through repetition suggests the creation of a *mise-en-abyme;* an imperious abyss that cocreated his issues with substance abuse. From Rex's example, we see a geographic and corporeal embeddedness of consciousness that is not necessarily about spatial enframements such as patriarchy or what it is to be a good father: men and women do family values, fathering, and community in much more complex ways than performance, identity, and representation allows.

The question of identity for Grosz and Bruno is focused on a virtuality that is, to use Deleuzian terms, simultaneously the future-past of the present. It is virtual also in the sense that it is vital. We are, precisely, what we do. For Grosz this is the force of geopower that has density and is a condition of existence. Put another way, we now have a new realm of images and experiences with an extended, and valid, reterritorialization of the so-called real that takes the abyssal contexts of the imperial and reworks them in tender ways. For Rex, this reworking moves him away from an imperious patriarchal form of fathering to something softer from where he can teach his boys.

While it is important to recognize that foundational norms within enunciations of identity are problematic, and while it is important to understand how the compulsion to repeat in an amnesiac spectacle resignifies again and again *en abyme* dominant discourses, it is important also not to foreclose inquiry into why and how particular identities emerge, their effects in time and space, and the role of subjects in accommodating or resisting fixed subject positions. We need to eschew methods and theories that reproduce a semi-Cartesian dislocation from geography, history, place, context, and antipatriarchal and anti-imperialist struggles. Tender mappings suggest a process of this kind.

Rex and I tried to create a map of tenderness that refocuses concern on the production of space and the degree to which ideology is inscribed in space and then acted out upon it and with it, while not missing the material and relational nuances of change, flexibility, freedom, and surprise that is the opening of political possibilities. It is clear that Western society has been characterized by an abstract spatial framing that is fragmented into subspaces devoted to the performance of specialized, homogeneous activities and to empire, but it is less clear how these may be transformed. The creation of ethnopoetic spaces may be one way of understanding the production of language and meaning along with the production of space as part of an anti-imperialist project. That language and space can be productive of reality is a primary focus of anti-imperialist projects. Rethinking identity, subjectivity, and space as fluid, relational, and inexhaustible does not foreclose upon the idea of a conscious, thinking—but not necessarily autonomous—subject. Nor does it foreclose upon new and different political possibilities.

CODA

February 19, 2011, was Rex's memorial service. The week before the service I got an email from his girlfriend, Angela, letting me know that Simon was not returning from Nebraska for the service. Angela was very upset by Simon's decision to stay away, and in her anguish she turned to the ethnopoetics that Rex and I had already putting together (figure 5.2):

> I just wanted to tell you that I re-read what you wrote about Rex in your book last night. Thank you so much for capturing his essence so perfectly. I felt like I was sitting with the two of you as he talked. It was immensely comforting and also very helpful in reminding me how important Manny and Simon were to him and how very much he loved them. I was so angry at Simon and very conflicted when he contacted me last week. Reading your book helped put things in perspective and helped me remember that Rex loved Simon in spite of everything and probably would want me to be kind to him.

Ethnopoetics are transformative to the extent that they are emotionally charged. They change lives by raising awareness that there are tender mappings beyond the abyss. In the first edition of the *Internationale Situationiste*, Debord calls for the study of how geography affects the emotions and behaviors of individuals.[35] This is precisely what Grosz meant by geopower, and it is accessible only through deep and tender mappings.

Appendix A. Ethnopoetics 1: Dialogue and Discussion

"And Manny stayed with you as you moved around?" I ask Rex.

"He came living with me when he was in fifth grade and by then I was living up by State College. [Both boys] used to see me a lot when I lived down at the beach 'cause both of them liked to surf. So, once I left the beach that was the last reason Simon had to come see me. And, you know, I mean..."

"Manny was in fifth grade, Simon would be...," I interrupt. It is a problem I have; my forgetfulness with numbers is countered by interruptions that interfere with the flow of dialogue.

"Manny was in fifth grade so Simon was in high school," replies Rex.

"So when you divorced Molly you moved back to the beach?" I have trouble with places also. Rex fills me in on his geographic solutions.

"First of all I moved to Mira Mesa. Okay, that was quick and handy. It was kind of close. I was working between Kearny Mesa and Escondido a lot and so it was a nice mid-point. I mean the I-15 traffic wasn't asinine yet. You know, you could do it. Within two years of Molly and I divorcing I got three DUIs, one of which was a felony. So I left... I put all my stuff in Mira Mesa in storage and went to see my mom and dad down in Kearny Mesa for about three weeks and then I worked a deal at work so that I took a partial leave of absence—all my vacation—and some excuse-time to do my jail term and then got to come back on because I could work when I went to work-furlough after I got out of jail. Then, after I got out of jail and work-furlough, that is when I went back down to the beach. Em, and the kids were by now up in Ramona. Their mom had sold the house and moved to Ramona. She'd re-married, and he was a very good guy. I have no complaints whatsoever. And he treats her good. Em, him and I actually get along very, very well."

"And Simon and Manny are living with the two of them?" I am still confused with the barrage of spaces and relocations.

"At this time. They are living with them as are his two kids from his first marriage."

"What time period are we talking about here?" Still confused over time.

"Oh, goodness." Rex seems a bit perplexed over timing these placements also. "So that would be ... I finally ... my last conviction was in 1983. Okay, so, and it was probably around, em, I am going to say 85/86 when Manny came to stay with me at State College. And he was in fifth grade. You know ... he was either right at the end of fifth grade or he was at the beginning because he started middle-school at Main ... over at Crawford. And he would have gone to Crawford. Em, Simon only came over to the place once and that was it. And I literally did not see him again until the week before he went into the military when he was pushing twenty-one. There are six years difference between the boys. And then Simon was gone in the military for nine years and after he came back, that is when we started re-connecting. Manny on the other hand. . . . It went . . . you know he kind of had the best of both worlds in terms of doing what he wanted when I was drinking 'cause I was drinking and I ended up. . . . He was getting a little too friendly on the streets and he knew which bars I hung at, so he would check in with me at the bars." Here is the memory and affect connection once more. Rex is quite clear about Simon visiting him only once before the estrangement that hurt him so much. So, Simon is out of Rex's life for a long time. I wanted to know more about how he connected with his other son.

"Sixth grade?" I ask, questioning at what age Manny was looking for his dad in the bars.

"Sixth and seventh grade. And in eight grade he got suspended a couple of times. And not realizing. . . . And I, you know, I mean what can I say ... I was an alcoholic father and I really wasn't attending to business." This part hurts almost as much as Simon's estrangements, but there is still an important, solid connection between Manny and his dad. Part of this connection, it turns out, is fomented by Manny's feelings over Rex's new wife.

"And what about your new wife?" I wonder where Susan is in all of this.

"Her and Manny didn't like each other and, em, interestingly enough, em ... Molly, the boy's mom, she couldn't stand her. And, eh, boy that one ended up in divorce. But that happened. The divorce happened when

I finally got sober. And, eh, so Manny... he was mad being up there stuck in the sticks 'cause there wasn't anything to do up there except you know work. And we had property, we had animals, and of course you know there was... it was up where [the] Viejas [casino] is now but when we bought up there Viejas wasn't there. There was a little market down the road so you had easy access to beer all the time and I drank at home. I didn't drink and drive any more. I didn't want to go to state prison. Em, but he, Manny... I encouraged him at school. I wouldn't work overtime on days when... 'cause living in Alpine he went to Granite Hills High School in El Cajon. So there was one bus down and one bus back. And if you didn't catch the buses you're on your own to get home. So he played basketball, he played Lacrosse, he did a lot of things down in El Cajon and he stayed out of trouble. He stayed out of trouble. And I would always pick him up on the way home, which was dicey sometimes."

Rex watches my puzzled look and decides I am not quite getting where he is going with this discussion of the trials that he is going through as a single-father of a teenage son. There is an aspect that he needs to return to.

"To back up a little, when Manny was eight he came down with type-one diabetes. So you know you always had to worry about his insulin levels and getting his insulin in him. So that had a few scary moments. So you know it was important to have, have people around him be aware. He almost died twice: diabetic comas. Both times [were] before we moved up to Alpine, and I have no idea why. You know. You know, at the time his Mom..." Rex pauses and deliberately stops where he is going with this line of thinking. When he starts talking again it is out of a different awareness, an awareness that fomented, perhaps, from the trials by space that had little to do with relocation and everything to do with reconsideration.

"It was all because of me."

"And you are separated from his Mom at this time?"

"Yeah, he's living with me. Yeah. Yeah. And we've talked about it since. Molly and I became very good friends again. You know, Manny told her I quit drinking and I quit drinking solely because Manny told me he couldn't take it anymore. And, you know, the wife at the time told me if I didn't quit she was going to leave. And I just said 'yeah, yeah.' You know the usual, I'll play that game 'cause I know she's going to drink and tweak before long. Which she did. But then when Manny told me, and there was something in me that I'd lost the relationship with my oldest boy and I could not—that was what made me quit drinking and learn to become sober. Because I couldn't lose both of them. That was my bottom, if you will. It certainly wasn't jail or fines or ER [emergency room] visits or any of that other stuff."

This was a huge life change for Rex, a reorientation that related solely to his connections to his boys. The pain of the possibility of losing another son was too much for him, and so he was pushed to find a solution that was not about geographic change but inner change. This was the crux of fathering for Rex, and I wanted to know more about the emotional precipice upon which his whole sense of self teetered.

"Tell me about that day Rex? Tell me the story of what happened when Manny gave you that ultimatum?"

"Well, he was in the living room. We were up in Alpine and I was probably out back, you know, taking care of the animals and the fruit trees or vegetable garden or playing with my dog. But I am sure I was drinking. I always drank. You know I mean that is what I did. So I came in the house, you know, for one reason or another. It could have been for anything, you know, to get another beer [or] to see how dinner was coming. I have no idea to this day why it was I came back in the house. And that is when Susan told me that if I didn't quit she was out of there. And eh, like I said, I had told her 'yeah, yeah, okay I'll quit.' Because I knew that wasn't going to have to stay permanent.

"And Manny just looked over . . . I'll never forget he had this look on his face like 'Dad, I can't do it anymore.' He just told me, he said, 'Dad if you don't I can't do it anymore.' And I said, 'Can't do what anymore?' And he says, 'I got to move back with mom, this is friggin' crazy.' He said, 'I never know what in the world is going to go on around here and you're going to be drunk, she's going to be doing something else. You know, I never know if she's going to go to the grocery store and buy food and come back with another horse.' Which she did several times, em, or another dog or a pig or, you know, whatever, a goat. We had 'em all. Which is okay, I didn't mind the animals. And, em, from him you know, looking back on it I mean this was a self-preservation move because he didn't know if there was going to be damn food in the refrigerator 'cause she'd be coming home with a horse. And I thought she was working, you know, I mean I am a good happy drunk making okay money and come to find out she was probably only working 10–15 hours a week and she was spending most of the time at the casinos. So, you know, it wasn't like she could stop and fill up the refrigerator with food, you know? And I remember telling him, I said point blank: 'Are you serious?' He said, 'Yeah, I'm serious.' I said, 'Okay, then I will.' And I said, 'Well what do you want me to do with all the beer?' and he said..."

At this point Rex start chuckling and his usually serious grey eyes start sparkling.

"He said... 'pour it out.' I said, 'Okay, help me pour it out.' We poured out about a case and a half of talls. I had one bottle of Jim Beam [that] we poured out... I didn't drink Jim Beam when Manny was around, I drank it on the weekends when he was up at his mom's because I promised him when we were back down at State College that I wouldn't drink Jim Beam when he was home anymore 'cause he usually had to put me into bed. And I couldn't do that to him anymore. So we poured it out and then.

"Now I have nothing to drink and I turned into an idiot. Angry. So she left anyway and now Manny and I deal with it. I mean I was on the edge. I was just going f-u-c-k-i-n-g c-r-a-z-y."

Rex says this slowly, lifting his fist and pumping it up and down for emphasis.

"I don't think I'd started going to AA yet. I started working more and all that. I tried the phony beers like Sharpe or O'Douls. I think those were the only two out and it took me about a 6-pack to realize that that was stupid. You know, I mean, what's the point? So Manny and I started doing things together and things were kind of going okay. And then I just flipped out. I just could not take not drinking. And so I just flipped out. It happened when I was at work and, and one of my co-workers got hold of the boss and the boss came over and took me down to the DAV [detox for veterans] program and I, you know, I mean I thought I was going to have to go to some 28-day lock-down place to get my head together or whatever. And how was I going to take care of Manny? And she told me she didn't think I needed it because by then I already had like a week and half, two weeks without any alcohol so she said you don't need to go to detox, you're already detoxed. And, eh, they got me into an outpatient program which was really, really good and really, really helpful. But the deal was I had to go there on Monday, Wednesday, Thursday nights and Saturday mornings, four hours each and plus I had to do four AA meetings a week. Well, I'm out in Alpine and that's down in Hillcrest, plus I am working out of Otay."

So much for Rex's geographic solutions.

"So I talked to Susan and she came home to take care of Manny while I was going through this."

"And you guys have split?"

"We've split. We were thinking of maybe getting back together, but no. As soon as I got done with the program she was gone. I filed for divorce. I got sober in October '91 she was gone by January '92. February, she was gone by February. And so it is just Manny and I up there and so I told Manny I said, 'okay, we're staying up here until you graduate.' He was a senior. He was going to be a senior. We had one more year and I said, 'When you graduate we are going back into the city.' And he said, 'okay,' and eh, Susan and I got divorced in May of that year, it was final.

"Manny and I stuck it out and we took care of each other and I didn't drink. I went to [AA] meetings and he was okay with that and he was datin' and doing his sports and you know we were really doing okay. Financially we were okay. Everything was okay, just that we weren't down in the city. [But things began changing] and, truthfully, em, I had to get rid of most of the animals, which was my joy of being up in the sticks. That and no neighbors: land, having animals and all that. But there was no way I could do it all [without Susan]. And I was ready to come back down into the city anyway, you know because I still had Manny and I am figuring he's going to be around for a while and I did want him eh. . . ." Rex pauses as his thought patterns slip latterly to some of the trauma that plagued his fathering at the time.

"Wow, it dawns on me my boy's diabetic and I don't want him driving up and down Highway 8 all hours of the day or night coming from school, coming from work, coming from friends of whatever. So, eh, I told Susan,—'cause both our names are on the thing, the mortgage—that I was going to put the house up for sale. And she said 'I'll buy you out,' which she sort of did. She took over the house when I moved out. And [Manny] started going to Grossmont, the Junior College. He was working full time."

"Where were you living?" I ask.

"Eh, we were living in Allied Gardens. Em, a found a house to rent over there close to school. You know, easy access to the freeway for me to go to work [and] for him to go to . . . he worked in Allied Gardens when I was still working in Otay. And eh, and it was okay. We were doing alright and em we both trusted each other. I chewed him out every now and again and he chewed me out every now and again if I did something stupid, you know, but, em, you know . . . both of my boys will tell you today that they learned a lot from me em, not just what not to do but a lot of what to do and when, and they also learned a lot from their mom and their step-dad. And both of them are good boys."

Rex's story is about movement, relocation, and the search for a place for his fathering. The rural idyll does not work because, for Rex, that geographic solution is not one about which he has control. His continued moves to find a better place to raise Manny certainly reflect his commitment—his trial by space—as a father, but it is the emotional trauma brought on by his son's threat of leaving that refracts the loss of Simon years before and highlights for Rex what is truly important.

Over the years and over many cups of coffee, Rex has relived with me that look on Manny's face, an emotional stab to his heart, a look that changed his life.

Now, years later, Rex's boys have moved on and have lives of their own. Manny is married with no kids and Simon is divorced and raising a son. And there is an important material space to the connection that he enjoys with his boys. It brings together his belief in what it means for him to be a father; it is the material practice of a father who wanted to teach. I ask Rex about his connection with Simon and his grandson and get an inclination of a connection that is primarily an emotional bond but is reified in Rex's idea of fatherhood. It is a fathering form that is material and sedentary.

"Simon and I are very, very close," he tells me, "we have been for eight and a half years or something like that and we talk all the time and at least once a week we talk on the phone and I talk to my grandson. Em, he has no problem at all with me being with his son. Em, and Simon has told me many, many things in the last five years when we really made a point to talk every week of different things he'd learnt from me while he was kid while I was still around that are with him today and some of them just blow me away. You know he used to sit there and look at my books and he'd say dad

I am just in awe that you've read all those books and that you can always find an answer for me."

Rex tells me a story about his fathering finding form. It happened when he lived with Molly and Simon was in grade school. Rex thinks of himself as an academic rebel, and he is proud that his sons also question the system.

"You know, when [Simon] was in grade school I used to, you know when they started to do book reports you know and the teachers in Poway—they were good people and all that but sometimes they did things. You know they would, they would have this preferred reading list that they wanted the book reports done on or a book of your choice at the bottom. Well, I would give him free reign to any of my books and eh and he was in trouble all the time because he'd never do book reports from the school district reading list and they were not topics that em, em the school district would prefer students reading. His very first book report was a book called *The Student as Nigger* ... caused an uproar. And I defended him to the hilt. He had to get his 'A' because it was an 'A' report, you know? And they told me to run my books through them and I told them, I told them when they were in charge of the thought-police I would consult with them. And ... buzz off, he'll read whatever he wants to read. And Simon has a very em, em a very good mind. He is very, very smart. He'll be graduated from college finally em this coming June with a 4.0 and he's already been accepted into em graduate school for MBA at Nebraska. And, he's come this far with 9 years in the army and raising a kid and, em, just pluggin' away."

"You must be proud," I say.

"I am. And, you know, one of the things that I regret with Simon was there was that huge period of fifteen years of disconnect—except if I heard something from his mom that was going on. And, em, he married and had one kid—Norm—interestingly he got married to a woman in the army who had a nine-month old kid and he raised that kid. Interestingly!"

What goes around, comes around, thinks Rex.

"Him and I talk about that now and he goes, 'Yeah, why would I? ... It was just the way it happened and I didn't think there was anything weird about it.' And I said, 'you were 9 months old when I met your mom.' And he goes, 'yes, she told me that later on.' 'Kind of strange son,' I said and he says, 'Strange what you can learn is a normal thing to do.' And the kids ended up, it wasn't his biological kid but Norm who is now 10, that is the one he is raising.

"So, the army sent him back to I think it is Fort Hood or Fort Bliss down there near El Paso, and promptly told him that he was going to have to do a two-year tour in Korea, and he just went ballistic. He said, 'you told me I'd been overseas too long, well guess what, I am not going!' And he got out of the army after nine and a half years. He went to work in LA at the Federal Penitentiary 'cause if he continued working for the feds his time in the army would count towards a pension. And she decided—his wife—she needed to go back to Nebraska where she was from, she couldn't deal with LA and all that stuff so she took the kids and left. And I heard something was going on and I called up there and he was all humming and hawing and I said, 'okay,' and this is like four o'clock on a Thursday afternoon or something like that and he wouldn't come out and say it so I just said, 'I'll talk to you later,' and eh . . . he lived in this apartment complex on the edge of East LA and eh, so I just filled up my truck and drove over there. It was in a security complex and I jumped the fence and got in and went and knocked on his door and asked, 'Okay what's happening?' Well, she was leaving."

"And this was when you and Simon started talking again?"

"Yeah, we started talking. It was at the very beginning of it. So they told me what was going on. She was leaving the next day, taking the kids with her and he was going to be alone and so I took him outside and we talked and I said, 'are you going to be okay?' and he goes 'Yeah, I gotta figure out how I am going to do it differently.' I said, 'Okay, whose helping you pack up all the furniture and stuff.' He said, 'I got nobody.' I said, 'I'll be back up tomorrow morning.' I drove back up to LA and helped her load the U-Haul with her stuff and I sat down and talked with him. And by now I've got some sober time on me and my concern is just helping him get through the deal at the time. Em, and that I think is what really reconnected Simon and I."

"You taking that action?" The doing that is fathering.

"Me just showing up, just taking that action in which . . . stuff that I would have done back before my alcohol got out of hand with not just my kids but any of my extended family: cousins, aunts, uncles. If they needed help we just showed up and did it."

"Yeah."

"Well, once I got to drinking I can't just show up. Either I don't remember that I'm supposed to be there or I'm incapable of getting there, you know? And, em, ever since then . . . He ended up moving back there and then, em. Ever since then . . ."

"Ever since when?"

"To Nebraska. . . . He worked with, he got a job with the DA (District Attorney) and em, last year was 2006 so 2005 Thanksgiving him and his wife end up getting divorced but they remained very good friends and their focus was on doing the best they could for Norm, which is Simon's biological son. The oldest boy is in a group home. He's . . . he's got some severe problems and em they not only couldn't afford the money to help him get into treatment or whatever, he was not safe to himself or others. So he's in Boy's Town I believe. Has been for a long time.

"So, she died. His wife died totally unexpectedly at Thanksgiving. She had her tonsils taken out and evidently had a, eh, severe allergic reaction to one of the medicines, antibiotics or something, and died. So, he is taking care of Norm and he's doing fine. And we talk at least once a week for an hour. I am going to go back and watch him walk when he gets his diploma. It is the right thing to do."

"Tell me some more about right things?" I ask. "You mentioned a couple of things that Simon talks about. What are a couple of things that you do right as a father that you're learning from your kids?" I want to know about what Rex has learnt about his fathering from his sons now that he is sober.

"I encouraged them to be honest." He replies. "Em, I encouraged them to do the right things for the right reasons knowing that that might not always be beneficial to them at the time they were doing it. I encouraged them both to go to college. Manny, em, one of these days he'll finish up. He has one year left at State College. I encouraged them to not go to college to look for an occupation. Go to get an education. Simon is going to get an MBA. What he would much rather get is a masters in philosophy or history. You know, he wants to teach philosophy or history, which just cracks me up because these are my two loves and I never tried to push my love of philosophy or history or political science or sociology or whatever on them. I wanted them to find their own area that they loved. Em, Manny met his wife over at State at a women's studies course he was taking. You know, he loved those classes. And he certainly wasn't looking for a wife at the time. That's for sure. And I encouraged them to find where . . . I said I don't care whatever it is you want to do in life to make a living I said I want you to find something that you like. And they said it was like my mantra and I would tell them all that—repeatedly—em there is nothing worse than having a job you hate because you will get up in the morning already hating the day. Don't get yourself in that bind. And I still tell them that. Now they hear me say it to Norm, my grandson. And he is like in fifth grade and they're going, 'Dad, you know?' And then three hours later you know they are telling him, 'listen to what your grandpa said about that job.' Norm is going to be okay. He is a very, very good boy. He is very bright. He's intelligent, artistic, speaks, like the arts. He's well rounded. He's got a lot of people back there in Nebraska that really, really love him and he is okay. You know, and he knows he's got em out here too and you know, I talk at long length with both the boys."

Rex and I talk some more about how he got into the counseling that he now enjoys so much. He is amazed at how much more closely connected are all parts of his life, and particularly his work and his relationships with his sons and grandson. It was not always this way. His past is one of finding solutions with constant moves around San Diego County, a job as an electrician on top of a telephone pole, and sons who did not trust him. Today, Rex senses a circularity between his core beliefs about what a father should be—a teacher and protector—and his lifestyle. He no longer fights to protect his sons by controlling their behaviors and remaining emotionally aloof. Rather, through letting go and taking care of his and his family's emotional needs, he finds freedom and connectivity. The context of his fathering is changing, and it has been a trial for Rex to come to terms with those changes. Reflecting on these changes he notes:

"When I first became a father, from the nine month mark on with Simon, I think back then fathers were expected to be two things. They were expected to be providers and enforcers, okay? Em, women were in the workplace but that was . . . the two income family wasn't necessarily the necessity that it is today, for most families. It is kind of like they did and then you had a couple of kids and you stayed home. Basically, you know, Molly and I could have bought any one of our houses, at the beach or the two houses we bought in Poway—qualified, paid for—and lived fine on my income. Wouldn't have had a lot of the extras but we would have done fine. Em I think fathering today is a different thing. I think fathers today em, rather than being an enforcer (I hate that word) but it is kind of—and there are still plenty of fathers out there today and plenty of women . . . you know, use corporeal punishment or whatever on a kid—I just disagree with that but you got to have a means to hold kids accountable and responsible for their actions that are appropriate for their age and size. But I think fathers today are trying their best as a rule to provide. I am not so sure that they are still trying to eh, eh, help guide, help educate. I don't see too many sitting around encouraging their kids to, to dabble in a lot of things.

"I think fathers today are falling into some bizarre trap, em, where it is more important that they be a friend to their kids than a father. And, you know, I don't mean you can't be you know a friend to your kids but your role really is to help prepare them for adulthood and most friends aren't going to help prepare you for your next step because they are just doing whatever you are doing with you, you know, I mean, it is a different thing and I know it is a difficult balance you know I did a lot of things: I used to go surfing and all that with my kids and I taught them how to swim and surf at early, early ages and be all kinds of different things but I think em I think part of society today has parents of both genders, but fathers in particular, running scared. You know I think there are fears . . . a lot of fathers today are afraid that they are going to get in trouble with the law, with CPS [Child Protective Services], with the school district, with whomever, if they put their foot down. And I don't mean they've got to beat the daylights out of their kid or anything like that but I think they've got them running scared. And I think kids today are being gypped out of parents you know, I really, really do. And part of that is a societal thing where it is all about me, don't you know?" Rex laughs. He knows that he is elaborating part of a patriarchal pattern here.

"You know, I think there are a lot of fathers out there today that are trying to come to terms with their past and they are afraid to face up to that past, whatever that past was. You know, em, and because they are afraid of that somehow the best abilities they have to be a father they can't really do that and then they get frustrated because they are not the fathers that they want to be. It is crazy. And I am not sure it was not the same way a hundred years ago. I am really not. That is my observation over the course of my life. You know, em, my dad did not want me to go off to war. I didn't want Simon to go off to war. I definitely didn't want him to re-enlist to keep going. . . . And when I felt that sense of relief knowing that Simon was out of Africa for at least then. . . . Craziest thing, the first thing I thought of was now I understand my Dad. And, you know, Simon was pretty cool, I mean he, he wouldn't, well he couldn't even tell his own family when he was going to do stuff and where, and when he'd be home. You know, but he was always somewhere or he was always plain clothes undercover in various parts of the mid-East, Europe, North Africa, South Asia. And, eh, he managed to live through it all. And he still wants to be a history and philosophy teacher. When he told me that I gave the phone to my girlfriend, and I did a cart-wheel in my own crippled way. And I am going, 'holy shit' you know?" Rex looks at me and there are tears in his eyes. This is the crux of his fathering; the unexpected that nonetheless ties into his core beliefs about fathering.

"And now he calls me: 'Hey Dad go to your bookshelf, and here's what I'm looking for.' and I go in there 'cause he knows I've got something on it with a reference list. Okay thanks, you know. And now it is, 'Dad when are you really going to get a computer so we can email?' And I promised him this year I'd get a computer. Ha!"

NOTES

1. Walter J. Benjamin, *The Work of Art in the Age of Its Technological Reproducibility and Writings on Media* (Cambridge, Mass.: The Belknap Press of Harvard University Press, 2008), 402.

2. David J. Bodenhamer, "The Potential of Spatial Humanities," in D. J. Bodenhamer, J. Corrigan, and T. M. Harris, eds., *The Spatial Humanities: GIS and the Future of Humanities Scholarship* (Bloomington: Indiana University Press, 2010), 26.

3. Michael Pearson and Michal Shanks, *Theatre/Archaeology: Disciplinary Dialogues* (London: Routledge, 2001), 64–65.

4. Paul Vidler, foreword to *Public Intimacy: Architecture and the Visual Arts,* by Giuliana Bruno (Cambridge, Mass.: MIT Press), 2007.

5. Diane Elam, *Feminism and Deconstruction: Ms. En Abyme* (New York: Routledge, 1994); Vidler, foreword.

6. Giuliana Bruno, *The Atlas of Emotion: Journeys in Art, Architecture, and Film* (London: Verso, 2002) and *Public Intimacy*. Elizabeth Grosz, *Chaos, Territory, Art: Deleuze and the Framing of the Earth* (New York: Columbia University Press, 2008).

7. Grosz, *Chaos, Territory, Art*, 18. Grosz does not use the term *geopower* in her 2008 book that brings together geography and art, Darwin and Deleuze. I first heard her use the term to describe a fundamental framing of the Earth at a panel discussion in 2012 at the annual meetings of the Association of American Geographers in New York.

8. Benjamin, *The Work of Art in the Age of Its Technological Reproducibility and Writings on Media*, 391.

9. Michael Rogin, "'Make My Day!': Spectacle as Amnesia in Imperial Politics," *Representations* 29 (Winter 1990), 99–123. Rogin terms the "spectacle of amnesia in imperial politics" as a conscious mirror of Guy Debord's *Society of the Spectacle* (Detroit: Black and Red, 1983/2000). For Debord, the current spectacle is precisely the transformation of desire and fantasy into the reality of the commodity occupying the totality of urban life. In an essay on the relations between dreams and desires to fantasies and commodity fetishism, I demonstrate the real and the fantasized city are inseparable. See Stuart C. Aitken "Dreams and Nightmares," in P. Hubbard, T. Hall, and J. Short, eds., *The Sage Companion to the City* (Los Angeles: Sage, 2008), 373–389.

10. Guilliana Bruno, *Streetwalking on a Ruined Map* (Princeton: Princeton University Press, 1992).

11. In actuality, there is no evidence that the military survey of Scotland, which took place in the ten years after the defeat of the Scots at Culloden Moor, was ever employed. It was never published, and it was only seen by the public in the last century as part of exhibitions. Rachel Hewitt, *Map of a Nation: A Biography of the Ordinance Survey* (London: Granta Books, 2010), 41.

12. Hewitt, *Map of a Nation*.

13. Stuart C. Aitken, "Opening Up the Hurt Locker," *Journal of Social and Cultural Geography* 13, no. 2 (2012), 117–133. Stuart C. Aitken "Leading Men to Violence and Creating Spaces for Their Emotions," *Gender, Place and Culture* 13, no. 5 (2006), 491–507; Stuart C. Aitken and Chris Lukinbeal, "Of Heroes, Fools and Fisher Kings: Cinematic Representations of Street Myths and Hysterical Males," in Nick Fyfe, ed., *Images of the Street* (London: Routledge, 1996), 141–159; Stuart C. Aiken and Chris Lukinbeal, "Mobility, Road Geographies and the Quagmire of Terra Infirma," in S. Cohen and I. R.Hark, eds. *Road Movies*, 349–370 (London: Routledge, 1997).

14. Henri Lefebvre, *The Production of Space* (Oxford: Blackwell, 1991), 23.

15. *Desiderius Erasmus, Novum Instrumentum ... Recognitum et Emendatum* (New Testament ... Revised and Improved) (Cambridge, 1516, printed by Johann Froben of Basel).

16. For an excellent geographical perspective on this uprising, see Oliver Froehling, "The Cyberspace 'War of Ink and Internet' in Chiapas, Mexico," *Geographical Review* 87 (1997): 291–307.

17. Rogin, "'Make My Day!,'" 100.

18. Desire for Freud and Lacan is about lack and repression, but for Deleuze it is about potential and is never repressive or bad.

19. Bruno, *Public Intimacy*, 163–165. The image may be found at http://arthistory.about.com/od/from_exhibitions/ig/Louise-Bourgeois/02-Louise-Bourgeois-Femme-Maison-1947.htm (accessed December 10, 2013).

20. Bruno, *Public Intimacy*, 165.

21. Bruno, *The Atlas of Emotion*, xi.

22. The image may be found at http://blog.art21.org/2011/03/21/weekly-roundup-95/louise-bourgeois-arch-of-hysteria-1993/ (accessed December 10, 2013).

23. Some of the work in this section is previously discussed in Stuart C. Aitken and Deborah P. Dixon, "Avarice and Tenderness in the Cinematic Landscapes of the American West," in Michael Dear, Jim Ketchum, Sarah Luria, and Doug Richardson, eds., *Geohumanities: Art, History, Text at the Edge of Place* (London: Routledge, 2011), 196–205.

24. Tom Conley, *Cartographic Cinema* (Minneapolis: University of Minnesota Press, 2009), 127.

25. Geoffrey Brereton, *A Short History of French Literature* (New York: Penguin, 1954), 116.

26. Pamela Cheek, *Sexual Antipodes: Enlightenment Globalization and the Placing of Sex* (Palo Alto, Calif.: Stanford University Press, 2003), 45.

27. Maria Walsh, "Film Space—Invisible Sculpture: Jane and Louise Wilson's Haptic Visuality," *Senses of Cinema*, no. 61 (2003), 3, http://www.sensesofcinema.com/ (accessed May 8, 2014).

28. Stuart C. Aitken, *The Awkward Spaces of Fathering* (Aldershot, U.K.: Ashgate, 2009) and *The Ethnopoetics of Space: Young People's Engagement, Activism and Aesthetics* (Aldershot, U.K.: Ashgate, 2014). A detailed account of the beginnings of ethnopoetry in the 1970s as part of anthropology may be found in the *The Ethnopoetics of Space*.

29. George Quasha, *Technicians of the Sacred: A Range of Poetries from Africa, America, Asia, Europe and Oceania*, rev. ed. (Berkeley: University of California Press, 1968).

30. Jerome Rothenberg, "Pre-face to a Symposium on Ethnopoetics." In Michel Benamou and Jerome Rothenberg, edrs., *Alcheringa/Ethnopoetics: A First International Symposium*, p. 6 (Boston Mass: Boston University, 1976).

31. In the 1970s, Heda Jason used the term *ethnopoetry* to describe and codify oral or folk literature from different societies in two different publications: *Ethnopoetics: A Multilingual Terminology* (Jerusalem: Israeli Ethnographic Society, 1975) and *Ethnopoetry: Form, Content, Function* (Bonn, Germany: Linguistica Biblica, 1977). By so doing, she attempted to isolate oral literature from its natural context in society and to break it up into its component parts. For Jason, ethnopoetry was understood as "verbal art, transmitted from generation to generation by talented performers in a process of improvisation" (*Ethnopoetry*, 5). She argued that a literary canon underlying ethnopoetic work enables the continuation of verbal performances. What I am doing here is quite different from the project Jason was engaged in. For example, with a project that discussed the multilingual terminology of ethnopoetics, Jason (in *Ethnopoetics*) attempted to define the discipline through its philological, anthropological, and semiotic roots. I want to distance myself from this mechanistic notion of ethnopoetry to use it, rather, as a methodology that embraces a consuming and lyrical basis for living and working. Thirty years later, ethnopoetry is no longer a subfield of anthropology. When used by academics it has become, simply, poetry about culture. Barbara Jones, for example, describes ethnopoetry as a "qualitative interprevist method of data analysis" (789) that uses poetry to stimulate discussion about data and to elicit emotional reactions and understandings of the work of pediatric oncology ("Tucked in My Heart: The Use of Ethnopoetry to Represent Meaning-Making of Social Workers in Pediatric Palliative Care," *Journal of Palliative Medicine* 9, no. 3 [2006], 789–790). She argues further that the method enables expression of complicated emotional experiences. Based upon a survey of 131 social workers, Jones creates a

fourteen-stanza poem written in the first person that reflects a myriad of emotions around the death of children. Similarly, Walt Nopalito Smith creates a long poem to help describe his bicultural experiences. He argues that a large part of this writing stems from his frustration with "scientific thinking" and the realization that "the reality in which I am participant-observing is *not* objective, but is emotion-*full* rather than emotion-*less*, and ... that reality is *not* separable from my own being" ("Ethno-poetry Notes," *International Journal of Qualitative Studies in Education* 15, no. 4 [2002], 461). For Smith, the process of ethnopoetry is also about "my own identity metamorphosis" that links his "creative power" to a move away from "the perceptual shackles of monocultural identity" and, in particular, the lens of scientific objectivity. Importantly, Smith points out that he is "still a lousy poet," but nonetheless, "cross-cultural poeting is downright good for the spirit of a constipated anthropologist" (462). I want to distance myself from this kind of solipsism as much as I want to get away from Jason's mechanistic methodology.

32. Rothenberg, "Pre-face to a Symposium on Ethnopoetics," 7.

33. Aitken, *The Awkward Spaces of Fathering*; Stuart C. Aitken, "Fathering and Faltering: 'Sorry, but You Don't Have the Necessary Accoutrements,'" *Environment and Planning A* 32, no. 4 (2000), 581–598; and Stuart C. Aitken, "The Awkward Spaces of Fathering," in Bettina van Hoven and Kathrin Hoerschelmann, eds., *Spaces of Masculinity* (New York: Routledge, 2005), 222–237.

34. Aitken, *The Awkward Spaces of Fathering*, 169–187.

35. Guy Debord, "Definitions," trans. Ken Knabb, *Internationale Situationniste* (Paris) 1 (June 1958).

6

DEEP MAPPING AND NEOGEOGRAPHY

BARNEY WARF

The construction and representation of difference, in whatever form, are core intellectual concerns of the humanities and social sciences. Acknowledging how different forms of knowledge are produced, their relations to daily life, identity, and social practice, and their epistemological implications are profoundly important. This theme is also a deeply geographic question, as knowledge is deeply embedded within particular spatial environments. The intertwined nature of knowledge and space is reflected in the widespread "spatial turn" across many disciplines.[1] In particular, over the last several decades, a renaissance in geographic information, propelled in part by the rise of new geospatial technologies and citizen-generated data, has markedly elevated the significant of knowledge of places.

This essay explores the intersections of two intellectual currents that recently have had powerful impacts on spatial thinking in the humanities and social sciences: deep mapping and neogeography. It opens with a brief definition of deep mapping as a means to represent places in all their untidiness and complexity. It holds that such representations of place, notably their unique qualities, have gained in importance in light of globalization, neoliberalism's ferocious impetus that pits places against one another, and tourism. Next, it points to neogeography as a means for constructing deep maps that allow the input of nonexperts and amateurs. Web 2.0 technologies have decisively transformed cartography, allowing locales to construct collective representations of place that suit their own purposes rather than those of academics or planners. Concerns of "truth" in this regard are often held secondary to their implications and consequences for particular epistemic communities. Third, the chapter asserts that neogeographic deep maps reflect and in turn sustain the contemporary epistemological shift into poststructuralism by accommodating a plethora of voices, diverse experiences, and worldviews, and by allowing such contending perspectives to be brought into a creative tension with one another. Fourth, it positions neogeography within the context of the digital divide and social and spatial inequalities in internet access. Finally, it offers an empirical example of just such a digital deep map from Brión, a town in Galicia, Spain, to point to the pragmatist orientation of such efforts.

DEEP MAPS: A BRIEF NOTE

Following William Least Heat-Moon's celebrated and exhaustively comprehensive portrait of Chase County, Kansas, *PrairyErth*,[2] which gave rise to the term *deep map,* this essay begins with the view that a deep map is a finely detailed depiction of a place, its history, landscape, and culture, and the people, animals, and objects that exist within it. Pearson and Shanks eloquently explain that "the deep map attempts to record and represent the grain and patina of place through juxtapositions and interpenetrations of the historical and the contemporary, the political and the poetic, the discursive and the sensual; the conflation of oral testimony, anthology, memoir, biography, natural history and everything you might ever want to say about a place."[3] Deep maps are thus inseparable from the contours and rhythms of everyday life, which falls squarely within the geographical tradition of time-geography and structuration theory that produced a long and rich intellectual history of theorizations about how place, biography, and social relations are interpenetrated.[4] Deep maps are not confined to the tangible or material, but include the discursive and ideological dimensions of place, the dreams, hopes, imaginations, and fears of residents—they are, in short, positioned between matter and meaning. Finally, such maps are also topological and relational in nature, subsuming each place's ties to other places, its embeddedness in networks that span spatial scales and range from the local to the global.

More recently, with the rise of critical cartography, deep mapping is afforded the opportunity to see such representations in avowedly political terms: the Foucauldian revolution in the social sciences and humanities has made it abundantly clear that discourses do not simply mirror the world, they also help to create it. As Denis Wood argues, maps help to bring into existence the world that they portray.[5] Thus, artist Clifford McLucas writes that:

> Deep maps will not seek the authority and objectivity of conventional cartography. They will be politicized, passionate, and partisan. They will involve negotiation and contestation over who and what is represented and how. They will give rise to debate about the documentation and portrayal of people and places. ... Deep maps will bring together the amateur and the professional, the artist and the scientist, the official and the unofficial, the national and the local. ... Deep maps will be unstable, fragile and temporary. They will be a conversation and not a statement.[6]

Deep maps differ from conventional maps in more ways than one. Deep maps not only show qualitatively more information than do shallow ones, they are self-conscious as social and political entities, making explicit their origins, purpose, whose interests they serve and whose they do not, and what they represent and what they do not. Deep mapping is thus far more than a technical exercise, but one imbued with powerful political implications aimed at changing the world they represent.

Every deep map reflects and contains more than one viewpoint. The diversity of voices, and whose are considered to be authoritative or "correct" (and whose are not) lies at the core of poststructuralist concerns with power, knowledge, and place. Poststructuralism is deeply suspicious of simple categorizations and master narratives that sweep complexity under the carpet: it serves as a sensitizing device to how representations are tied to interests and embedded in webs of power and pays particular attention to the voices of the marginalized, the subaltern, and powerless.[7] In lieu of portraits of places as coherent, integrated, and tidy, poststructuralism offers a range of embodied perspectives often at odds with one another and celebrates the inchoate chaos that undermines any aspiration of imposing a single narrative structure on a complex and heterogeneous reality.

NEOGEOGRAPHY, DEEP MAPS, AND
POSTSTRUCTURALIST EPISTEMOLOGIES

The explosive growth of digital technologies, as well as the interactive capacities commonly labeled Web 2.0, has important methodological and epistemological implications for deep mapping. Online, interactive mapping facilitates a bottom-up reconfiguration in how data are collected, transmitted, analyzed, visualized, and utilized that differs considerably from traditional top-down models in which experts and government agencies dictate the criteria of data collection, analysis, applications, and standards of truth. The set of spatial representations and practices that emanate from Web 2.0—neogeography—includes an enormous volume and variety of representations that are highly contextual, personal, and relevant only to the producers and consumers of these data, not to academic experts.

At the core of the shift into neogeography is Web 2.0, the set of software-revolutionizing applications of the web. Key to the architecture of this technology are asynchronous Javascript and XML and application programming interfaces, which facilitate the creation of websites similar to desktop applications while allowing instantaneous user interactions. The functionality offered by Web 2.0 has precipitated significant changes from traditional approaches: whereas traditional cartography and geographic information systems (GIS) are expert-centric, neogeography is obviously user-centric. With the spread of Web 2.0, neogeography, code spaces, and the automatic production of space have become deeply woven into the fabric of daily life for countless millions of people to such an extent that simple dichotomies such as "online" and "off-line" fail to do justice to the ways in which the real and the virtual worlds are profoundly interpenetrated.[8] In this sense, neogeography has fostered an unprecedented democratization of geographic knowledge, often with roots far removed from academic experts. Thus Goodchild maintains that whereas "the early Web was primarily one-directional, allowing a large number of users to view the contents of a comparatively small number of sites, the new Web 2.0 is a bi-directional collaboration in which users are able to interact with and provide information to central sites, and to see that information collated and made available to others."[9]

Neogeography refers to the process whereby people using an eclectic set of online geospatial tools to describe and document aspects of their lives, society, or environment in terms that are meaningful to them.[10] A broader interpretation of neogeography includes the study of cultural mapping in all realms of everyday life, catalyzed by the digital mapping technologies and social networking practices of Web 2.0. It has also been greatly accelerated by the explosive proliferation of mobile digital technologies, including smart phones and location-based services. The interactive websites characteristic of Web 2.0 allow users to upload data about locations and apply them in diverse ways, including, for example, simple displays of locations (e.g., favored bird watching sites) or lists of attributes of a place near a user equipped with a global positioning system. This approach lies at the heart of mapping websites such as Google Maps, Yahoo!Maps, OpenStreetMap, and Bing Live Maps. Unlike traditional approaches, users can create share and use information via "crowd sourcing," which allows large, widely distributed groups of people to work together toward a common goal.[11] For example, Jackson reported that OpenStreetMap attracts two hundred thousand volunteers annually.[12] Google Maps was used by 71.5 million people in 2007 and Google Earth by 22.7 million.[13] Rather than rely on state- or corporate-produced data, neogeography generates volunteered data/content, relocating the center of knowledge production from a handful of self-appointed experts to large numbers of people with limited formal geographic training. Sui labels these changes the "wikification of GIS," after Wikipedia, the famously popular, user-generated, online encyclopedia.[14] Rana and Joliveau argue that to fully appreciate the significance of this phenomenon, neogeography must be regarded "as an extension of mainstream geography for everyone made by everyone."[15] As Goodchild points out, neogeography often implies that academic geography is redundant or unnecessary; for this reason, he advocates instead terms such as volunteered geographic information.[16]

Neogeography has important epistemological as well as pragmatic implications.[17] Because neogeographical knowledge emanates from personal interactions with places, what is held to be true is often highly contextual and specific to the community of users who generate and use such data. Maps in this context do not aspire to be "objective" depictions but are tailored to the needs and outlooks of narrow communities of interest. By rejecting the all-seeing Cartesian subject as their epistemological reference point, critical deep maps demonstrate and help to popularize the social and spatial embeddedness and embodiment of *all* perspectives. The analytical focus is bottom-up, that is, on the relations among large numbers of actors rather than individuals. Through the properties of emergence or self-organization, complexity theory implies that local spatial configurations of interactions affect outcomes at broader systemic scales. Neogeographies are, in a sense, also complex adaptive systems in which conscious agents construct and reproduce systems of meaning, both intended and unintended. Neogeographic practices do not follow a trajectory of gradual, linear growth, but indeed often burst into existence suddenly when large numbers of followers adopt them quickly, a theme consistent with the discoveries of the emerging network science.[18]

Neogeographic deep mapping is an ontological counterpoint for poststructuralist epistemology in that it allows for multiple voices to be heard, leading to a cacophony of representations in which places are depicted and viewed through multiple lenses. Moreover, in that it allows people to jump vast distances instantaneously, cyberspace is both a metaphor of relational space and identity—as fluid, open-ended, produced and reproduced through inter-relations, and forever coming into being—and simultaneously a vehicle to understand people and places in those terms.[19] As the internet has become indispensable to ever-larger numbers of people, neogeography charts paths through the virtual metaverse and mirror worlds, the multiple spaces and "digi-places" it contains. Neogeographic spaces, generated through countless bottom-up interactions of users who are widely dispersed among many physical locations, are virtual, constantly changing, and often bear only tenuous linkages to material geographies. In this sense, neogeographic space is compatible with the Deleuze/Guattarian "flat ontology," or in geographic terms, spatialities uncontaminated by the obfuscating effects of scale, as demonstrated by the multiple, multifaceted examples in the book *Else/Where: Mapping New Cartographies of Networks and Territories*.[20]

Neogeographic deep maps have profound philosophical implications for how knowledge is constructed and the criteria by which it is deemed appropriate, or not. For example, groups utilizing neogeographic technologies are not likely to generate random samples of data, a criterion that lies at the heart of commonly accepted definitions of "the scientific method." How reliable are the results of neogeographic approaches without random sampling, the lynchpin of scientific respectability?[21] As diverse groups of people with varying agendas harness Web 2.0 to upload their own data and interpretations, conventional views of what constitutes valid knowledge and truth come into question. What, then, is "true" when people, particularly nonexperts, generate their own data and stories to interpret and make sense of the world in ways that may be markedly at odds with the rigid criteria demanded by academic experts? Furthermore, user-generated content also tends to rely on very polarizing samples—only those who either love or hate the subject matter beyond a threshold level tend to post their opinions online. These extremes can drown the mainstream interpretations and generate conceptual and political discord.[22] In neogeography, the sharp divisions between knower and known, representations and the world they portray, epistemology and ontology, are deliberately blurred, and truths (for there can be more than one) are repositioned as a partial, contingent series of statements that reflect lived reality and are useful in it. This line of thought owes much to John Dewey's and particularly William James's pragmatist epistemology, in which "truth" is determined and confirmed by its utility and effectiveness in application, that is, from its consequences; thus, "the 'true' is only the expedient in our way of thinking, just as the 'right' is only the expedient in our way of behaving."[23] In this line of thought, there is no single observable reality or one approach methodological style: rather, there are multiple realities, many different and equally viable ways of constructing knowledge, all of which require an enormous tolerance for inconsistency, incompleteness, and uncertainty.

As Jürgen Habermas argues, communications are central to the social process of truth construction, through which individuals and communities of interest partake in the public, discursive interpretation of reality.[24] Habermas's "ideal speech situation" consisting of unfettered discourse is central to the "public sphere" in which social life is reproduced and through which truth is constructed in the absence of barriers to communication. Truth in this reading is inseparable from lived experience, intent, and social practice, leading to the consensus rather than correspondence theory of truth. In this reading, all participants in a debate would theoretically have equal rights and abilities to make their views known and to challenge any other view; when all power relations have been removed from the freedom to engage in discourse, the only criteria for resolving contesting claims is their truth-value. And, importantly, "the participants in an ideal speech situation [must] be motivated solely by the desire to reach a consensus about the truth of statements and the validity of norms."[25] If one adopts a Habermasian approach in which democracy is approximated by an "ideal speech situation" of unfettered discourse in the public sphere, neogeography provides a reasonably good approximation. Of course, access to neogeography technologies is limited by social factors such as income, education, and often ethnicity and gender, all of which are significant determinants of the "digital divide" as well.[26] Nonetheless, inasmuch as anyone with simple access to web tools to upload data and download results can participate in neogeographic communities, Web 2.0 essentially realizes Habermas's ideal speech situation for vast pools of participants.

Richard Rorty offered a famous shift in metaphors for understanding knowledge in light of the collapse of the Cartesian world view.[27] If the mirror served as the ideal "reflection" of knowledge construction from the Enlightenment onward, with its emphasis on vision, accuracy, and light, then the conversation serves as the perfect vehicle to summarize the forms of poststructuralist knowledge construction in the present era: a messy series of dialogues in which each voice is partial, incomplete, and contingent. Neogeography is thus much more of a "conversation" than it is a "mirror" of the world, in which the truths constructed are relative and useful for specific communities, there is no value-free arbiter to decide what is accurate (true or not), and accuracy is decided by consensus and pragmatic value. Moving away from an ocularcentric view of truth that lies at the heart of the world-as-exhibition, the conversational view of truth puts more emphasis on practice and performance, on the speech acts that lead to points of agreement between contending worldviews. Neogeography has moved us closer to the performative theory of truth with emphasis on actual spatial practices, that is, "performing" these geographical or cartographic tasks on a sustained, repeated basis. Instead of truth, "performativity" or "truthiness" has been pushed to the front, that is, the quality of an idea "being done right" or "feeling true" without voluminous evidence.[28]

NEOGEOGRAPHY AND THE DIGITAL DIVIDE

A central concern to the pragmatic viability of neogeography is the digital divide in internet access. Inequalities in this regard are manifested at multiple spatial scales. While roughly 2.3 billion people used the internet worldwide in 2013, or roughly 32 percent of the planet, there are enormous discrepancies in access among and within countries. Everywhere, internet access is greater in cities than in rural areas, and internet usage is higher among the young, well educated, and well off; frequently ethnic minorities, and often women, have markedly lower rates of use.[29] For many people—the familiar litany of the poor, the undereducated, ethnic minorities, and the socially marginalized—the internet remains a distant, ambiguous world. Denied regular access to cyberspace by the inability to purchase a personal computer, the technical skills necessary to log on, or public policies that assume their needs will be magically addressed by the market, information have-nots are deprived of many of the essential skills necessary for a successful or convenient life. While those with regular and reliable access to the internet drown in a surplus of information—much of it superfluous, irrelevant, or unnecessary—those with limited access have difficulty comprehending the opportunities it offers, the savings in time and money it allows, and the sheer convenience, entertainment value, and ability to acquire data from bus schedules to recipes to global news. What is more, within most countries, internet usage is concentrated within cities, whereas in much of the developing world, vast numbers of people, often the majority of the population, live in rural areas; in such places that typically suffer from slow connections, graphical information, which uses much more bandwidth than text, is virtually out of the question.

As the uses and applications of the internet have multiplied, the costs sustained by those denied access rise accordingly. At precisely the historical moment that contemporary capitalism has come to rely upon digital technologies to an unprecedented extent, large pools of the economically disenfranchised are shut off from cyberspace. As the internet erodes the monopolistic roles once played by the telephone and television, and as the upgrading of required skill levels steadily render information technology skills necessary even for lower wage service jobs, lack of access to cyberspace becomes increasingly detrimental to social mobility. Indeed, those excluded from the internet may be more vulnerable to social forces they do not and often cannot perceive than ever before.

Given these discrepancies, and the resulting uneven abilities to participate in neogeography, utopian claims that neogeographies lead to a "flat world" (e.g., Thomas Friedman) must be viewed with considerable skepticism. Contrary to the hyperbole that continues to swarm around the internet, multiplying even faster than do viruses and webpages, cyberspace reflects all of the inequalities and social divisions that permeate the nonvirtual world. Far from constituting some mythologized world of unfettered individualism, as some advocates portrayed it, cyberspace in fact is thoroughly shot through with relations of class, gender, ethnicity, and other social categories.

Despite these limitations, neogeographies offer great promise in allowing people in various parts of the world to represent themselves in their own terms. Digital deep mapping is much more than scratching an aesthetic itch, it has important political repercussions. In an age of rampant globalization, in which capital's ability to pit places against one another has reached new heights, the differences among places assume renewed significance. In contrast with popular mythology that holds that globalization homogenizes places, in fact global processes must be tailored to the specifics of individual locations. This theme is critical in understanding the importance of place promotion as localities vie with each other in attempting to attract capital and tourists. Deep mapping via neogeography therefore offers places an unprecedented ability to portray themselves as their residents wish them to be portrayed, not simply how local elites or firms wish the place to be represented.

AN APPLIED NEOGEOGRAPHIC DEEP MAP: BRIÓN, GALICIA

An existing neogeographic deep map of a place can be found in the case of Brión, located in Galicia, in northwestern Spain. Brión is a small town (population seven thousand) with an unusually extended web presence. Its deep map was started in 2003 in an initiative of faculty at the nearby University of Santiago de Compostela called SINDUR (Sociedad de la Información y Desarrollo Urbano-Regional, The Information Society and Urban-regional Development).[30] Its website was developed over several years in the early 2000s as a collective effort that involved significant input from local residents, many of whom were unfamiliar with digital technologies, particularly Web 2.0. The project was instigated, for free, as part of a national and regional effort to overcome the limitations of Spain's digital divide and to harness the power of the web for marginalized communities. In some respects, the community served as an experiment to understand the implications of the wider transformation into an information-based economy and society. This initiative received enthusiastic support from the municipal government and arose in conjunction with nationally subsidized attempts to promote other uses of cyberspace such as online banking, e-commerce, distance learning, and to widen public access to information about government services. Brion's interactive webpages were greatly enabled by the implementation of free Wi-Fi services in its public spaces, including schools, parks, and libraries. As a result, internet access became very widespread, reaching almost universal coverage. Local primary and secondary school teachers played key roles in promoting interest in the project among the town's young people. Many residents contributed using their mobile or cellular phones, while others relied on the expertise of local teachers, librarians, and government staff.

While many communities have representations on the web, Brión's digital deep map is unusually extensive, including the following: a plethora of pictures and films; interviews with residents; personal biographies and life stories; samples of local music; depictions of the region's wildlife and natural landscapes; historical accounts of the town's past; an online museum; paintings; architectural details; interviews with elected officials and business leaders; portraits of sports teams and events, folk culture, and emigration; and interactive opportunities for lessons and marketing for small local firms.[31] As of 2012, its website contained more than two hundred individual reports by residents, three thousand pictures, five hundred songs, and two hundred videos. This is not a view that attempts to represent the town coherently as a neatly packaged destination for tourists, but grapples with its people, past, and landscapes in all their contradictions and messy complexity. It is arguably a web equivalent of *PrairyErth*.

Thousands of uploads to the Infobrion website were posted by a diverse array of the town's residents over several years. Some, particularly the elderly, were unfamiliar with the internet, and a few refused to learn how to participate. Many entries are amateurish and of relatively low technical quality. Residents insisted that the site's text be in the local Gallegan language rather than Castillian Spanish. Crucially, the website's contents and organization were determined by the collective preferences of the town's residents, and their priorities often surprised the academic experts who facilitated the project. Some people published diaries of their day-to-day activities, yielding a rich trove of time-geographic information about the contours of daily life. Residents compiled a database of more than 110,000 events in the town's history. One popular application was a documentation of the town's folk culture as it evolved over the years. Unexpectedly, an online school of horse riding emerged, complemented by myriad representations of equine festivals and contests. The local music school started online courses in music appreciation, with videos of local performances by students. The effort also gave rise to an interactive local atlas listing local residences and points of interests; maps were based on uploads of data on sixteen thousand locations in the town and neighboring areas. An online, interactive GIS allows users to query these places and make their own maps. Some residents used the site to commemorate loved ones who had died or left the town; others recorded oral histories of the very old: in a community with a disproportionately large number of elderly, this was no trivial matter. Local farmers and gardeners became enthusiastic proponents of the project as way to show off their skills with crops. One offshoot was Granxa Familiar, a network used to advertise local produce in other towns and villages. For Galician nationalists, the site became a way to celebrate the region's Celtic heritage, including, for example, the local festival of Samain (which in medieval Britain became Halloween). Exploits of the local *futbol* (soccer) team were described in great detail, including boasts of victories over rival teams from nearby towns. An online ethnography museum included photos of local handicrafts, notably shoe and hat production, as well as paintings and songs. One section contained biographies of emigrants from the town, some of whom had returned after years abroad. A virtual lost-and-found was complemented by a local version of eBay. Candidates for local political offices used it for their campaigns, and the government posted election results there. Others came to use the site as a means of keeping track of news, announcements, and lists of upcoming events or for Facebook-like personal webpages. Because the town's population is not homogeneous in outlook, occasionally disputes erupted over the "correct" representation of some events, particularly those with historical and political implications, and the site's chat rooms served as a public forum for airing these differences.

The InfoBrión project gives us a glimpse of what a neogeographic multimedia deep map can look like. With initial inspiration from academics and support by government officials, the website took on a life of its own. Far from being dominated by the views or interests of elite experts who ostensibly view the town through the eyes of a disembodied Cartesian subject, the mass of information uploaded to the site, which is constantly in flux, reflects the needs and interests of the town's inhabitants. While most of the material was local in content, some explored the community's connections to other places, as with emigration or sports rivalries. Because this material was intimately related to the town's residents' bodies and lives, concerns of "accuracy" or scientific integrity here are displaced by the criteria of relevance and utility: it is in many respects an epistemological pragmatist's utopia. InfoBrión illustrates applied neogeography in practice and substantive differences from traditional planners. Local planners were often reduced to listening to residents' views and were unable to enforce designs constructed in the antiseptic environment of planning offices. Rather, the town's self-image and priorities emerged organically as an emergent property and were far more democratic as a result.

CONCLUDING THOUGHTS

In an age of explosive globalization, spatial knowledge has become more, not less, important. Deep maps are important vehicles for mobilizing local collective consciousness and giving form to geographical imaginations. By putting the sources of data and the means to visualize it firmly in the hands of users and consumers rather than those of an elite group of expert producers, neogeography forces a broader recognition of the degree to which truth-values reflect broader social concerns such as trust, reputation, and credibility.[32] Accuracy, in this context, is largely a matter of ideology and preference, contingent upon context and purpose, and tailored to the specific interests of communities of interest. In facilitating the emergence of numerous "neighborhoods of truth," neogeography encourages us to abandon the holy grail of universal generalizations and come to terms with the place-bound nature of geographic knowledge. In its stead, it opens up venues for viewing the world through the eyes of particular groups bound together by lifestyle, political values, recreational habits, and other dimensions of social life. However, consideration of how neogeographies are made must pay due attention to the digital divide and attendant inequalities in access to cyberspace: for those who cannot log in, neogeographies are little more than an entertaining fantasy.

The InfoBrión project described here offers a useful example of a neogeographic deep map in practice. The literature on this topic, and cyberspace in general, is often top-heavy with ponderous theoretical concerns and light on examples. The empirical application of neogeographic technologies by communities, as opposed to individuals, remains largely ignored. In Brión, aggressive efforts to confront Spain's digital divide, particularly the lagging usage of the internet in rural areas and small towns, led to a sustained attempt to create a robust local digital community. As a large share of the town's population became involved, the InfoBrión website came to reflect their interests and priorities, which were often at odds with the detached views of academics and experts. The collective effort of Brión's residents to construct a digital representation of their lives, history, culture, and community served their purposes, as it should. While the results ranged from the intriguing to the trivial, it is important that such efforts be evaluated in terms of their utility and consequences. In short, whether viewed as an exemplar of William James's pragmatism or a Habermasian ideal speech situation, this example illustrates the profound epistemological as well as utilitarian implications of digital deep mapping. "Truth," whatever that may be, or knowledge in general never float in some aspatial, asocial void, but can only exist when laced through the contours of everyday life and meaning.

NOTES

1. B. Warf and S. Arias, eds., *The Spatial Turn: Interdisciplinary Perspectives* (London: Routledge, 2008).
2. W. Least Heat-Moon, *PrairyErth: A Deep Map* (New York: Mariner Books, 1999).
3. M. Pearson and M. Shanks, *Theatre/Archaeology* (London: Routledge, 2001), 64.
4. A. Pred, "Place as Historically Contingent Process: Structuration and the Time-geography of Becoming Places," *Annals of the Association of American Geographers* 74 (1984), 279–297.

5. D. Wood, *The Power of Maps* (New York: Guilford, 1992); D. Wood, *Rethinking the Power of Maps* (New York: Guilford, 2010).

6. C. McLucas, "Deep Mapping," http://documents.stanford.edu/MichaelShanks/51 (last accessed October 24, 2013).

7. J. Williams, *Understanding Poststructuralism* (Montreal: McGill-Queens University Press, 2006).

8. N. Thrift and S. French, "The Automatic Production of Space," *Transactions of the Institute of British Geographers* 27 (2002), 309–335; M. Dodge and R. Kitchin, "Code and the Transduction of Space," *Annals of the Association of American Geographers* 95 (2005), 162–180; E. Gordon, "The Geography of Virtual Worlds: An Introduction," *Space and Culture* 11 (2008), 200–203.

9. M. Goodchild, "Citizens as Sensors: The World of Volunteered Geography," *GeoJournal* 69 (2007), 211–221.

10. A. Hudson-Smith, A. Crooks, M. Gibin, R. Milton, and M. Batty, "NeoGeography and Web 2.0: Concepts, Tools and Applications," *Journal of Location Based Services* 3 (2009), 118–145.

11. J. Howe, "The Rise of Crowdsourcing," *Wired*, 14 (2006), 1–4, http://www.wired.com/wired/archive/14.06/crowds.html (last accessed October 24, 2013); J. Howe, "Crowdsourcing: Why the Power of the Crowd Is Driving the Future of Business," 2008, http://www.slideshare.net/ajturner/how-neogeography-killed-gis?src=embed (last accessed October 24, 2013).

12. J. Jackson J., "Openstreetmap Attracts 200,000 Volunteers," http://www.pcworld.com/printable/article/id,186096/printable.html (last accessed October 30, 2013).

13. M. Haklay, A. Singleton, and C. Parker, "Web Mapping 2.0: The Neogeography of the Geoweb," *Geography Compass* 2 (2008), 2011–2039.

14. D. Sui, "The Wikification of GIS and Its Consequences: Or Angelina Jolie's New Tattoo and the Future of GIS," *Computers, Environment and Urban Systems* 32 (2008), 1–5.

15. S. Rana and T. Joliveau, "Neogeography: An Extension of Mainstream Geography for Everyone Made by Everyone?" *Journal of Location Based Services* 3 (2009), 75–81.

16. M. Goodchild, "NeoGeography and the Nature of Geographic Expertise," *Journal of Location Based Services* 3 (2009), 82–96.

17. B. Warf and D. Sui, "From GIS to Neogeography: Ontological Implications and Theories of Truth," *Annals of GIScience* 26, no. 4 (2010), 197–209.

18. M. Batty, *Cities and Complexity: Understanding Cities with Cellular Automata, Agent-based Models, and Fractals* (Cambridge, Mass.: The MIT Press, 2007).

19. D. Massey, *For Space* (London: Sage, 2005); J. Murdoch, *Poststructuralist Geography: A Guide to Relational Space* (London: Sage, 2005).

20. S. Marston, J. P. Jones III, and K. Woodward, "Human Geography without Scale," *Transactions of the Institute of British Geographers* 30 (2005), 416–432; J. Abrams and P. Hall, eds., *Else/Where: Mapping—New Cartographies of Networks and Territories* (Minneapolis: University of Minnesota Press, 2006).

21. A. Bruns, *Blogs, Wikipedia, Second Life, and Beyond: From Production to Produsage* (New York: Peter Lang, 2008).

22. C. Sunstein, *Going to Extremes: How Like Minds Unite and Divide* (New York: Oxford University Press, 2009).

23. W. James, *Pragmatism and the Theory of Truth* (Cambridge, Mass.: Harvard University Press, 1907/1978), 2.

24. J. Habermas, *The Structural Transformation of the Public Sphere* (Cambridge, Mass.: The MIT Press, 1989); C. Calhoun, *Habermas and the Public Sphere* (Cambridge, Mass.: The MIT Press, 1992).

25. J. Bernstein, *Recovering Ethical Life: Jürgen Habermas and the Future of Critical Theory* (New York: Routledge, 1995).

26. M. Crang, T. Crosbie, and S. Graham, "Variable Geometries of Connection: Urban Digital Divides and the Uses of Information Technology," *Urban Studies* 43, no. 13 (2006), 2551–2570.

27. R. Rorty, *Philosophy and the Mirror of Nature* (Princeton: Princeton University Press, 1979).

28. F. Manjoo, *True Enough: Learning To Live in a Post-Fact Society* (New York: Wiley, 2008).

29. B. Mills and B. Whitacre, "Understanding the Non-metropolitan–Metropolitan Digital Divide," *Growth and Change* 34, no. 2 (2003), 219–243; L. Kvasny and M. Keil, "The Challenges of Redressing the Digital Divide: A Tale of Two US Cities," *Information Systems Journal* 16 (2006), 23–53; P. Poncet and R. Blandine, "Fractured Space: A Geographical Reflection on the Digital Divide," *Geojournal* 68, no. 1 (2007), 19–29.

30. C. Sexto, X. Arce, F. Quintá, and Y. Vázquez, "Alfabetización Digital en Comunidades Marginadasa Partir de un SIG. Estudio de caso en Galicia," *Anales de Geografia* 29, no. 1 (2009), 223–234.

31. Brión's digital deep map is available at www.infobrion.com. Last accessed on May 7, 2014.

32. M. Bishr and L. Mantelas, "A Trust and Reputation Model for Filtering and Classifying Knowledge about Urban Growth," *GeoJournal* 72 (2008), 229–237; M. Bishr and W. Kuhn, "Geospatial Information Bottom-up: A Matter of Trust and Semantics," in in S. Fabrikant and M. Wachowicz, eds., *The European Information Society* (Berlin: Springer, 2007), 365–387; A. Flanagin and M. Metzger, "The Credibility of Volunteered Geographic Information," *GeoJournal* 72 (2008), 137–148.

7

SPATIALIZING AND ANALYZING DIGITAL TEXTS

Corpora, GIS, and Places

IAN GREGORY, DAVID COOPER, ANDREW HARDIE, AND PAUL RAYSON

INTRODUCTION

Once upon a time not so very long ago, it was all very simple—information technology (IT) was concerned with storing and analyzing databases of numbers. The discipline of statistics—which predated computing by centuries—provided suitable techniques for taking the large amount of numbers held in a database and summarizing them and the relationships between them, using a much smaller number of summary statistics and graphics. Thus the use of IT involved quantitative data and social science approaches, and, conversely, if you did not use quantitative sources or were suspicious of social science approaches you would not use IT.[1] Developments over the past decade or so have meant that this cozy dichotomy of mutually incompatible approaches is breaking down. Suddenly, and in many ways without much fanfare, IT has become primarily concerned with text. Through both "born digital" sources, such as email, the worldwide web, and social networking, and digitizing paper sources to create digital libraries and digital archives that have the potential to contain every book ever written, IT has undergone a fundamental shift. Today the bulk of the content that is created for IT is, in fact, text. This presents a major challenge. While statistics was well suited to analyzing large numeric databases, there is no similar discipline for text. The disciplines that place the study of texts at their center are those in the humanities. Unlike statistics, humanities disciplines have traditionally eschewed approaches that quickly summarize large amounts of content, and instead stress the importance of using reading—close reading—to understand the subtleties and nuances within the text. While this approach will rightly remain the gold standard for understanding texts, it has one fundamental flaw: it is far too slow to be the only approach to understanding large bodies of text in a world where the researcher has access to literally billions of words of content. This results in humanities researchers having to be highly selective, and this tends to be done in a way that is far more arbitrary than most humanities researchers would like to admit.

These developments might be thought to place the humanities at a crossroads, where researchers are faced with the choice of carrying straight on and continuing to read texts in a detailed but slow way, thus failing to exploit much of the content that is available, or alternatively taking a sharp turn to new methods that summarize millions or billions of words without the researcher ever having to read any of them. The reality is more complex and more subtle: reading will remain central to humanities-based approaches, but it needs to be enhanced and complemented by methodologies that exploit the digital nature of modern text. These methods will enable us to summarize large corpora which would be difficult, if not impossible, to read in their entirety. By extension, this process will allow us to determine which parts of the corpora we ought to read closely (and, conversely, which we should not), thereby helping us to justify textual selectivity. It will also help us to understand how what is read closely relates to the broader context of the remaining unread material within the corpus. Such methods, then, will necessarily involve a shuttling between distance and proximity, abstraction and particularity: processes which correspond, at least in part, with Franco Moretti's controversial model of "distant reading."[2] At the same time, the adoption of such methods will also facilitate self-reflexive thinking on the processes through which we, as academic researchers, select those texts which become the objects of detailed critical analysis; thinking which, in turn, will raise further questions about the nature of canon formations. Finally, such methods will also allow texts to be placed within networks which stretch beyond the circumscribed boundaries of the digitized corpus: they will allow texts from a disparate range of sources, and in a diverse range of genres, to be brought together in a single analysis; and they will permit a range of other forms of data—including, for example, statistics, images, multimedia formats, and, of course, maps—to be integrated within the same digital space.

This chapter argues that geographical text analysis—the combination of geographical information systems (GIS) and corpus and computational linguistics—offers one way to implement this. A GIS is a combination of a database and a computer mapping system in which every item of data is *georeferenced* to give it a real-world location.[3] This structure offers a number of advantages: it allows the researcher to explore the database by location to ask questions such as What is here? and What is near here?; it allows data to be mapped to summarize the geographies that the database contains; it allows data from different sources to be integrated because all data are underlain by real-world coordinates; and it provides a platform for spatial analysis, a form of statistical analysis in which the locations of the items under study are explicitly included with the analysis.[4] GIS has traditionally been a quantitative technology whose use within the humanities has been restricted mainly to the social science ends of history,[5] although there are increasing calls for GIS to be more generally applicable within the humanities.[6] Corpus linguistics is a methodology used to study language using a large naturally occurring body of text—a *corpus*—on a variety of levels including lexis, syntax, semantics, and pragmatics or discourse. It has been employed, notably, for lexicography or dictionary creation, where large corpora are used as source data for determining sense boundaries, definitions, and examples for dictionary entries. Corpus techniques are increasingly being exploited across a wide range of areas within linguistics, such as the description of grammars, the analysis of literary style, or the investigation of language change. As a preliminary step in many corpus-based techniques, automatic language analysis techniques from the closely related area of computational linguistics, otherwise known as natural language processing (NLP), are used to enhance the corpus data with some sort of annotation to code one or more levels of the analysis in a robust and consistent manner.[7]

Bringing these fields together provides an approach that offers the potential to develop spatial narratives that tell us how different places are represented in different ways. Extending this will also provide information on how representations of different places have changed over time and among authors, genres, and/or publications. It can also allow this information to be integrated with georeferenced statistics, maps, or images to see whether the different sources are telling the same or different stories. As yet, we are in the early stages of developing such techniques; however, pilot work has been encouraging.[8] This chapter will describe the ways in which place-names can be automatically extracted from a corpus using NLP techniques so that they can then be linked to a gazetteer to georeference them and read them into a GIS. Once this database creation phase is complete, the next stage is to analyze these data. Here two entirely different forms of analysis need to be brought together: those from GIS and spatial analysis concerned with the analysis of geographical patterns, on the one hand, and those from corpus linguistics and NLP concerned with analyzing texts, on the other. Finally we look at some of the broader implications that this approach could have to humanities disciplines. The chapter is based on the early stages of a European Research Council–funded project Spatial Humanities: Texts, GIS, Places, which is developing suitable techniques to implement these approaches and applying them to two separate studies, one of which will create a *literary GIS* of the English Lake District, while the second will focus on nineteenth-century social history by integrating quantitative and qualitative approaches.[9]

CREATING AND SEARCHING A CORPUS

A computer corpus[10] may be created in many ways; one central concept in corpus linguistics is that the researcher's corpus must be well suited to the research question that they wish to study. In the early days of corpus linguistics, the only way of creating machine-readable text was to manually type in all the texts to be included in the corpus. Nowadays, manual transcription can be avoided for some sorts of text by employing optical character recognition (OCR) software to extract machine-readable data from images scanned from the printed page, though some types of text—such as spoken texts—still need to be typed in. An increasingly large proportion of the texts we might wish to include in corpora of the contemporary language are now "born digital," especially those available via the web. Historical corpus data is typically among the kinds of data where typing (or else extensive manual correction of OCR output) is required. Although the methods described in this chapter rely on full text sources being available and are more accurate when the quality of the OCR or transcription is high, there are significant efforts and initiatives underway aimed at improving the quality of OCR in historical and literary texts and at undertaking large-scale full-text transcription.[11] However the text is produced, a standardized style of markup based on XML may be applied to the corpus to indicate the structure of each text, although it is also possible to apply corpus methods to a "raw" corpus without any markup.

The first step in moving from a raw text corpus to a spatial database is rooted in one of the foundational techniques of corpus analysis, namely concordance searches. A *concordance* is a data output consisting of all matches in a corpus for a specified search pattern, plus a specified amount of the text surrounding each match, the *cotext*. This technique permits a researcher to begin to engage with the text by reading the cotext and thus carrying out qualitative analysis in order to draw meaningful conclusions about the occurrences of the words or phrases searched for. Concordance-creating software usually permits the researcher to extend each concordance line to provide more surrounding context, or to jump straight into the full text at that position in a manner similar to a hyperlink on a website. An example of a concordance of the place-name *Stirling*, from the Lancaster Newsbooks Corpus,[12] a corpus of news books published in London in the seventeenth century, is shown in table 7.1.

All corpus search software allow the corpus to be searched for a specified word or phrase, which can often include wildcard symbols. Using more advanced tools, we can also search a corpus for all the words that have been given a certain tag in the annotation process. When we wish to extract information about the places discussed within a corpus for the purpose of building a GIS, our starting point is a concordance. Critically, we do not just want to find all the different place-names. We want to find all the different *instances* (or "mentions") of all the place-names. If *London* is mentioned eight thousand times and *Foston on the Wolds* is mentioned once in a million words of text, then the list we create needs to contain eight thousand instances of *London* and one instance of *Foston on the Wolds*. This way the list accurately reflects the relative prominence of different place-names in the data.

Table 7.1. An example of a concordance based on *Stirling* in the Lancaster Newsbooks Corpus

TEXT	CONCORDANCE		
DutchDiurl5	The General intends next week to march towards	**Stirling**	, and so for the hills. The Lord Craighall
FScout156	to be their Commander in chief, their coming near	**Stirling**	, and col. Lilburne's forcing them to the Hills
FScout156	(though the weather was unreasonable) to march from	**Stirling**	against them; but as he appeared, they quitted
FScout175	divers soldiers out of Leith, and about 50 from	**Stirling**	are gone to the Hills; from whence they descend
FScout178	it is certified, That Gen. Monk is advancing towards	**Stirling**	, and intends to cut his passage through the Hills,
FScout180	, whereby he signified, That he was advancing beyond	**Stirling**	towards the Highlanders from whence he intends,
PerfAcc175	companies are come up. The General will be at	**Stirling**	next week. From Milford Haven, May 8. All
PerfAcc175	Scotland do advertise, that General Monk was lately in	**Stirling**	, and is now on his advance towards the
PerfAcc176	General marcheth tomorrow from Dalkeith towards	**Stirling**	, and from thence to some of our frontier
PerfAcc176	yesterday removed his Headquarters from Dalkeith to	**Stirling**	from whence he intends, after securing some new
PerfAcc177	, provisions being so exceeding scarce in those parts.	**Stirling**	May 16. Thursday last the General came hither
PerfAcc177	to and from the hills, about 12 miles from	**Stirling**	, General Monk is marched that way to observe
PerfDiOc02	that General Monk is upon his march on this side	**Stirling**	, to join with us, his coming may prove very

Note: The text column on the left-hand side provides information on which text the reference was taken from.

The most straightforward way to get an exportable concordance of place-names is to search for a part-of-speech tag for *proper nouns*. Proper noun is the grammatical category of names of people, places, organizations, and other specific entities, which often behaves differently to *common nouns,* names of categories of entities. In English, proper nouns are typically capitalized whereas common nouns are not. Most grammatical corpus annotation will be able to identify and mark up proper nouns—for example, the CLAWS (Constituent Likelihood Automatic Word-tagging System) part-of-speech tagger developed at Lancaster University[13] marks proper nouns with the tag NP1. A concordance for NP1 will therefore extract all mentions of all place-names that have been correctly tagged as proper nouns (it is possible, of course, for the tagger to make mistakes). It will also extract all mentions of names of persons or organizations but, as is discussed later, these can be filtered out at a later stage.

An alternative approach would be to use NLP techniques for *named entity extraction.* These techniques allow the automatic discovery of names of people, places, organizations, times, dates, and other quantities, using criteria other than simply spotting grammatical markers of proper nouns. The accuracy of such techniques is quite variable and depends on the subject and quality of the input text. For example, if one has a list of all major cities in the United Kingdom stored as a gazetteer then it is quite straightforward to find all mentions of such place-names in a large quantity of newspaper text with simple fuzzy matching techniques. However, to find all names of people and distinguish them from place-names and other organizational names is more problematic. For example, *Lancaster* (a city) would have to be distinguished from *Lancaster Bomber* (a plane or a type of beer), *Stuart Lancaster* (the England rugby coach), and *Duke of Lancaster* (a nobleman or a pub) by using phrasal patterns and possibly further context in order to decide on the meaning. As with other NLP problems, a variety of approaches including knowledge-based (dictionary lookup), rule-based (hand-crafted templates), and statistical methods (using probabilities and trained language models) exist, and these may need to be tailored for a particular domain or input text. The problem of disambiguating our initially extracted list of instances is one that we must address—either at the initial stage, using these automatic named entity extraction techniques, or manually at the georeferencing stage. In either case, we must draw on features of the cotext to decide which mentions become part of our dataset for georeferencing.

FROM A CORPUS TO A GIS

Using corpus techniques thus provides a list of all of the suspected place-names within the corpus. Converting these into a GIS is conceptually simple; however, there are a number of practical difficulties that may make this difficult and time-consuming. The core of the process is to join the list of suspected place-names to one or more gazetteers to provide a location to georeference each place. A gazetteer is a database table that provides a coordinate for each place-name that it contains. In its simplest form it has three columns: place-name, x-coordinate (or longitude), and y-coordinate (or latitude). Some gazetteers also hold additional information such as which higher-level administrative unit the place lies within, what type of feature it is, and different variations of the name's spelling.[14] A number of gazetteers are available including Geonames,[15] World-Gazetteer,[16] the Getty Thesaurus of Geographical Names,[17] and the Ordnance Survey's 1:50,000 Gazetteer.[18] In theory, joining the list of suspected place-names to the gazetteer using a relational join will give coordinates to all of the suspected place-names and, from here, it is simple to read these coordinates into a GIS software package where they will form a point layer.[19] In practice, the process is more complicated as a result of a number of practical problems that can lead to three types of errors: names that are not place-names being wrongly allocated a coordinate; real place-names not being given a coordinate; and place-names being allocated to the wrong coordinate.

The first of these occurs when a suspected place-name, such as *Lancaster*, is not being used as a place-name. Errors of this type can be filtered out using the cotext in which the suspected place-name occurs using the same phrase-patterns that named entity extraction often exploits. Thus, if an instance of *Lancaster* is preceded by *Mr., Duke of* or a proper noun known to be a person's forename, then it can be removed from the list. In the same way, if it is followed by words such as *bomber,* these can also be removed. This is the first way in which the cotext of each instance of a suspect place-name, extracted via the concordance search, is crucial for the transference to a GIS. It could also be argued that occurrences of the suspected place-name that are preceded by *the* or *a* should be removed, as in "the Lancaster flew many missions" or "a Lancaster was sighted"; one might also remove plurals such as "Lancasters were involved in the raid on." However, if filters like these are used too mechanically, there is a danger of removing place-names that may be wanted. There may also be genuine questions about whether some words should be included as place-names or not. *A Lancaster graduate* clearly refers to the university and thus the town and arguably should be included as a place-name, but it is equally true that this is not a direct reference to the town. As a consequence, filtering cannot be applied across the board as a standardized procedure for all lists of suspected place-names, but is instead a crucial step in the research process that involves making judgments about what does, and does not, constitute a reference to a place-name. This judgment can only be made by exploring the context in which the suspected place-names are used.

The second type of potential error, where suspected place-names fail to match to the gazetteer, are conceptually simpler. They can occur because the place-names' spellings differ between the text and the gazetteer; because the place-name is not in the gazetteer; or because the suspected place-name is genuinely not a place and should be omitted. Spelling variations are very common, particularly in historical sources, although they persist today. The town of *Saint Helens* can be spelt with at least three possible variants of *saint* (Saint, St., or St) and *Helens* may or may not have an apostrophe. This gives a total of six possible variations, excluding spelling mistakes and digitizing errors. When variants are found, whether they are genuine differences or spelling mistakes, they can be added to the gazetteer along with a standardized version of the spelling. Similarly, if places have been omitted from the gazetteer, these can also be added to it. In this way, the process of georeferencing a text also becomes a way of improving and enhancing gazetteers.

The third problem is places being allocated to the wrong coordinate. This usually occurs because there are two or more places with the same name and the gazetteer either has the wrong version or contains several places with the same name. In the second case, disambiguation can be done manually, or automated procedures can be developed to decide which option is most likely to be correct.[20] The first option is more difficult and only careful checking is likely to spot problems of this sort.

It should be clear from this that the process of georeferencing a list of suspected place-names requires a certain amount of manual intervention and even then is still error prone. Careful checking can reduce these errors significantly, but requires a considerable investment of time. Even if this can be invested, if a large corpus is to be georeferenced the results will still contain errors, and while these can be minimized, subsequent analyses should be sensitive to this.

ANALYZING GEOREFERENCED DIGITAL TEXTS

Once place-names have been identified and georeferenced, they can subsequently be analyzed using both corpus-based approaches, which have traditionally been primarily used by linguists, and GIS-based approaches, which have traditionally been used by geographers. Crossing this divide has much to offer to the methodologies used by both disciplines and, more importantly, much new knowledge to contribute in the humanities and social sciences. As briefly described in the introduction, the intention is to be able to summarize large corpora and to highlight texts, or sections of texts, which are considered worthy of more detailed analysis. We have already seen the role that the concordance technique plays in building a corpus-derived GIS. A further technique can be used to enhance the analysis: *frequency profiles.* Frequency profiling of counts of place-names—or other words—identified within a text can be used to highlight variation within one text or a collection of texts. An example is shown in table 7.2, which counts the number of instances of proper nouns containing the string "ford." Additional information on where within the texts these instances are found would allow us, for example, to highlight sections of a text where high concentrations of certain place-names occur. By further processing the resulting frequency lists, we may also be able to spot groups of place-names that regularly (or never) occur together. Using a corpus comparison technique called *keywords,* it is also possible to highlight which place-names occur more often than expected in one section of the corpus relative to the whole corpus or to another reference dataset.

While these techniques allow us to ask *where* a corpus is talking about, and how this changes, the question that is likely to be of more interest is the following: What does the text say about these places? To begin to answer this, we need to engage with two further automated techniques

Table 7.2. An example of a frequency profile for proper nouns containing the string "ford" in the Lancaster Newsbooks Corpus

WORDFORM	TAG	FREQUENCY
Oxford	NP1	38
Milford	NP1	28
Hereford	NP1	19
Bedford	NP1	18
Stafford	NP1	18
Seaford	NP1	14
Crawford	NP1	11
Hertfordshire	NP1	11

Note: Proper nouns are tagged as NP1. Only words that occur more than ten times have been included.

from the corpus researcher's toolbox: *collocation* and *semantic analysis*. A collocation analysis addresses co-occurrence—which words or phrases regularly occur together in a text or corpus. This is a quantitative abstraction and summation of the cotext. For example, a reading of the concordance for *Stirling* shown in table 7.1 quickly and clearly illustrates that the town was largely being talked about in relation to its military significance. Collocation analysis allows us to identify such patterns without the need to read every concordance line, although examination of at least some concrete examples by the researcher is usually necessary. Since collocations are a key indicator of semantics, lexicographers can use them to help differentiate and cluster word meanings. Collocation can be used comparatively by corpus linguists to discover differences between the representation of words and concepts in certain genres. For example, we may compare the concepts of *bachelor* and *spinster* in romance fiction, or research the representation of *immigrants* and *asylum seekers* in the U.K. press.[21] When exploring georeferenced texts, collocations of place-names can be used to discover what words writers use in the text surrounding the occurrences of those place-names. At some level of aggregation, this may permit overall patterns of description to emerge from a text or collection of texts that would otherwise not be available from a simple reading of those texts. Such patterns may show what modifiers (adjectives) are being used to describe a particular place or what activities (verbs) occur close by. By combining the collocations extracted with a qualitative analysis using concordances, a researcher is able to form categories from groups of co-occurring words in the context of place-names and start to highlight the key descriptions that emerge from the underlying texts.

In fact, we can automate this grouping of terms to some extent. As described earlier, an automatic part-of-speech tagger can be used to mark up proper nouns in text. Using a *semantic* tagger, we can assign to the text another level of tags representing semantic fields of words and phrases. The semantic field represents each word or phrase's position within a general ontology. Each semantic tag groups words from a dictionary together into similar categories of meaning, for example, education (P1), warfare (G3), and farming (H4). The tagger used here is the UCREL (University Centre for Computer Corpus Research on Language) Semantic Analysis System, which incorporates a dictionary of over seventy-five thousand words and phrasal templates designed by hand, in combination with a small number of other rule-based and statistical techniques to assist in the selection of the correct semantic tag in context.[22] Although the accuracy is high (91 percent),[23] there are inevitably some meanings that the system does not know or that it will tag incorrectly in specific domains, and these need to be considered in any subsequent analysis. Using the UCREL Semantic Analysis System in combination with a collocation analysis allows us to analyze patterns of semantic tags that regularly co-occur in the text near to each place-name. This therefore assists in the analysis of the *concepts* associated with a particular place, addressing questions such as: Is that place associated with themes such as education or warfare? Or, more generally, is it presented in a positive or negative way? With enough occurrences of a specific place-name within a text or corpus, we will start to see patterns emerging through the collocation analysis, to show statistically significant relationships between the place-name and a concept or group of concepts. These will then need to be confirmed through a qualitative analysis using close reading of concordance lines or extended extracts from texts in order to check for errors from the automated processing. As outlined in the following sections, the collocation analysis allows different kinds of mappings to be generated from the same overall list of mentions, for example, mappings associated with different concepts.

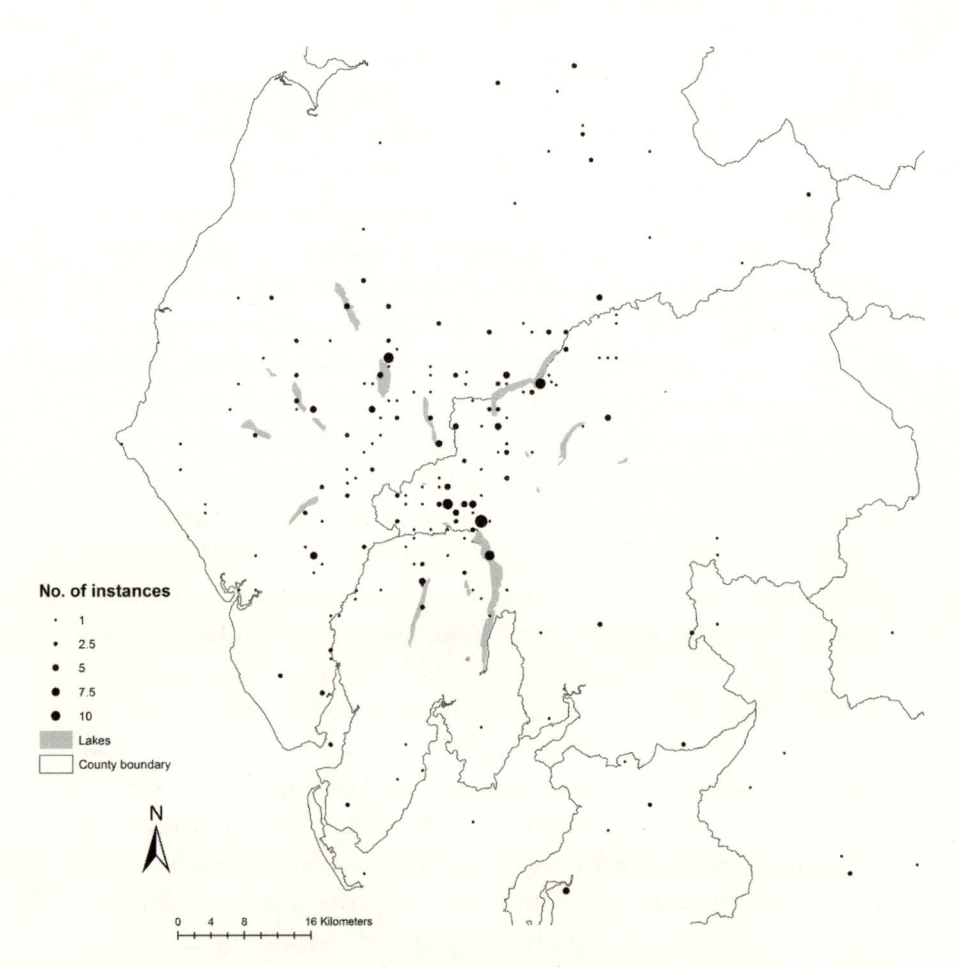

FIGURE 7.1. Proportional circles representing place-name instances from a small corpus of writing by William Wordsworth.

GIS-BASED APPROACHES

GIS and spatial analysis techniques offer a completely different, and radically new, method of exploring texts. The easiest way of "analyzing" a georeferenced text is simply to map the points that are named within the text. This can be done using a simple point map of the places mentioned or perhaps using more complex symbology to show for example: which source the places were named in; the date they were mentioned; or their collocates, for example, whether they are talked about in association with a certain theme or certain type of emotional response. An example of this, taken from a small corpus of material by William Wordsworth, is shown in figure 7.1. This is very straightforward within a GIS and presents a simple but effective overview. From here there are two very different routes for further analysis: *map-based querying* and *spatial analysis*.

In map-based querying, an interactive map is used that allows the text or texts to be explored through a map-based interface using technology such as Google Maps or Google Earth.[24] Figure 7.2 shows an example of this based on a small corpus of Lake District literature where a user has clicked on the point on the map representing Sca Fell. In the top-right of the screen this has returned a concordance of all of the mentions of Sca Fell (including spelling variations), and this, in turn, has been used to find a particular mention in a text shown beneath the concordance. Clicking other place-names in the text would, in turn, highlight these on the map. This approach allows the reader to ask the question What has been said about this location? and then read all of the responses in the corpus. It is thus well suited to close reading but makes use of the hyper-textuality within digital texts such that rather than reading in a linear manner from beginning to middle to end, the reader can switch from place to place within and between texts.[25] In this case, geographic location provides the hyper-textual structure through which the reader can approach the analysis (or reading) of the text. This, in turn, leads to the possibility of creating distinctly spatial narratives as space becomes the prime organizing motif behind the way in which the texts are organized.

While map-based querying provides an approach that fits well with the humanities tradition, spatial analysis comes firmly from the social and Earth sciences. While this means that the approaches offered must be used sensitively, it does not mean that they should be rejected, as they provide a highly effective way of summarizing large and complex geographical patterns. One highly effective technique, originating from criminology and epidemiology, is *density smoothing*.[26] This accepts that maps of point patterns are difficult for the human eye to comprehend, and that the points tend to suggest a level of precision that place-names within a text cannot support. Instead, density smoothing creates a continuous surface with high values being found in areas with many points. This provides an effective way of summarizing either all of the places

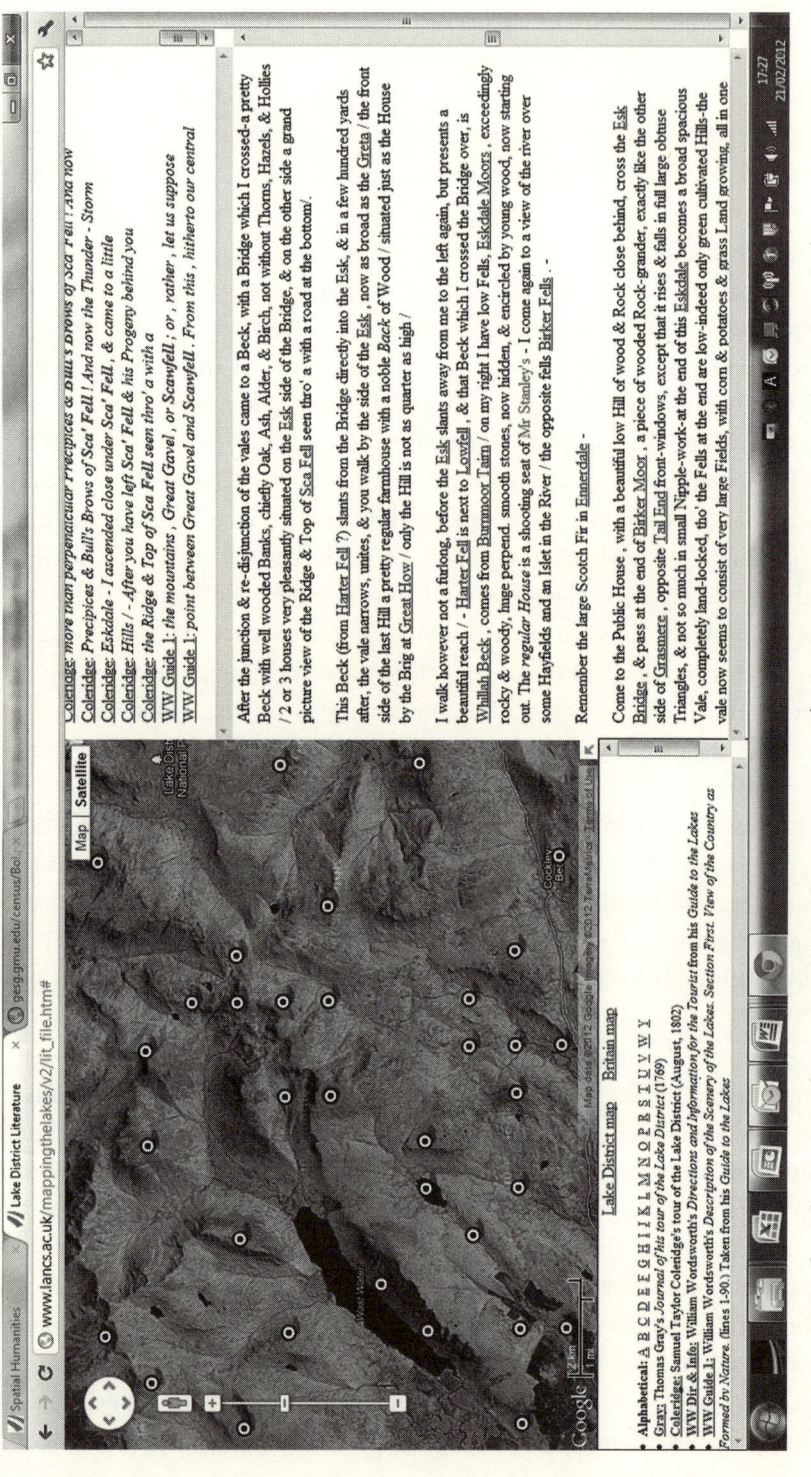

FIGURE 7.2. Map-based querying of a corpus of Lake District literature (http://www.lancs.ac.uk/mappingthelakes/v2).

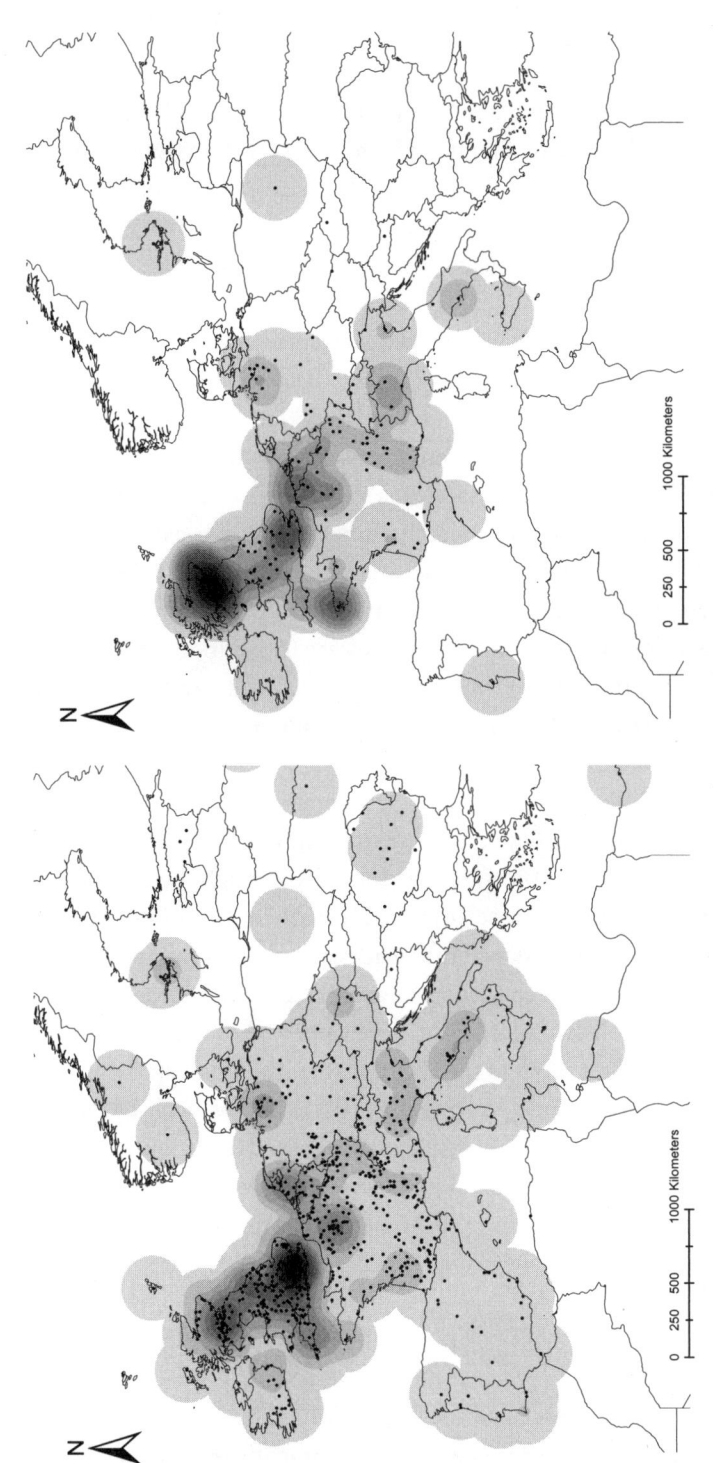

FIGURE 7.3. Density smoothing of place-names mentioned in the Lancaster Newsbooks Corpus showing (*left*) all places mentioned and (*right*) places mentioned that collocate with words related to war (tagged as G3). A point indicates one or more mentions while the shading shows the density of mentions with darker shading indicating more mentions in an area. (Derived from Gregory and Hardie, "Visual GISting," figures 3 and 5.)

mentioned in a text or, when used in combination with techniques such as collocation and semantic tagging, place-names that are found near to words with specific meanings. Figure 7.3A shows an example of this based on mapping *all* of the places mentioned in the Lancaster Newsbooks Corpus, while figure 7.3B shows places that collocate with words tagged as being associated with war.[27]

While previous work has shown density smoothing to be a highly effective technique for summarizing texts,[28] it is only one of a number of well-established spatial analysis techniques that are used to explore point patterns to see if they cluster, are evenly distributed, or are randomly distributed.[29] Techniques such as Moran's I, Geary's G, and local indicators of spatial autocorrelation[30] would all be suitable for exploring this. Developing these ideas further, spatial analysis techniques could be used to explore whether places associated with certain words or themes cluster near other words or themes or whether they are found in totally different locations. In spatial analysis terms, this requires a multivariate technique such as multivariate local indicators of spatial autocorrelation or possibly geographically weighted regression.[31] A slightly different question where spatial analysis will also be able to help concerns whether the pattern of place-names found throughout the texts follow a logical progression such as moving around a study area sequentially as might be expected to be the case in conventional travel literature. Finally, an additional component can also be added to ask whether places and their associated themes change over time, or between authors or genres. All of these questions can be asked using established spatial statistical approaches with some minor modifications to adapt them to the use of texts rather than statistics.

From a technical perspective, therefore, there are two well-developed fields that can be used to analyze texts: corpus-based techniques and spatial analysis. To date these have had little to do with each other; bringing them together will provide an exciting new potential to understand the geographical patterns and meanings within texts.

IMPLICATIONS FOR HUMANITIES: THE EVOLUTION OF THE LITERARY GIS

So far this chapter has described the ways in which texts can be analyzed geographically. While these are becoming technically feasible, the key issue that remains is how useful they are to applied scholarship within and beyond the spatial humanities. In considering this, we focus on how such methodologies can facilitate the spatial analyses formulated by literary critics—scholars whose work has traditionally been predicated upon the close reading and qualitative interpretation of a relatively small number of texts. By extension, how can such exploratory digital humanities research feed off, and back into, contemporary theoretical debates regarding the literature of space, place, and landscape and, in particular, the emergence of geocritical practice as codified by Bertrand Westphal? The potential and problems of a literary GIS were first explored, however, in the Mapping the Lakes project.[32] This project used place-name georeferencing to map out the spatial narratives of two key Lake District topographical prose texts: Thomas Gray's account of his 1769 touristic tour of the region, and Samuel Taylor Coleridge's documentation of a characteristically singular walking excursion through the western half of the Lakes in August 1802.[33] This initial attempt to construct a deliberately delimited literary GIS generated both technical and critical findings. From a technological perspective, the project circumnavigated the presentational problems traditionally associated with codex-based reader-generated literary cartographies through the direct linking of electronic texts and digital maps. At the same time, Mapping the Lakes transcended the limitations of problematically positivist applications of GIS through the mapping of the emotional geographies to be traced in the respective primary texts: a form of textual mood mapping which has clear correspondences with "sentiment analysis" methodologies and which intersects with the wider multidisciplinary evolution of qualitative GIS research.[34] Close textual analysis was used to identify those locations at which Gray and Coleridge articulated positive, negative, and even inherently contradictory experiences of the Cumbrian landscape and environment. As a result, Mapping the Lakes negotiated a rapprochement between the use of GIS—a digital tool almost invariably associated with the visualization of significant quantities of spatial data—and the nuanced close reading of textual textures which defines much literary critical practice.

The Spatial Humanities: Texts, GIS, Places project has begun to build upon this initial development of literary GIS through the digitization and georeferencing of, among other sources, a large corpus of Lake District landscape writing, including guide books, regional histories, autobiographies, journals, and poetry. One of the primary aims of this part of the project is to produce an electronic archive which will bring together, in a single digital space, a significant number of a heterogeneous range of topographical texts written between 1750 and 1900. This process is leading to the creation of a fully accessible scholarly resource: a corpus which allows the user to identify the richly intertextual nature of both canonical and historically marginalized Lake District literary texts. This electronic archive will not remain absolutely fixed, though, but instead will exist as a textual environment which can be continually developed through the digitization of further place-specific texts.[35] The digital intertext, therefore, is characterized by a state of openness and will be subjected to ongoing growth and expansion. Crucially, as the texts are georeferenced, this facilitates the visualization of the large-scale geographical patterns which are embedded in texts written during a key period in the region's spatial history. The GIS highlights those sites which emerged as central to the dominant spatial narratives of this particular landscape: those fells, villages, lakes, and pathways which were subjected to multiple layers of textual representation and re-representation. Alongside this, the GIS also draws attention, through its blank spaces, to those locations which were placed on the edges of the region's hegemonic cultural geography. This is a form of digital humanities practice, therefore, in which the corpus and the GIS are brought together to produce what might be described as a spatial intertext of this culturally overdetermined terrain.

The development of this methodology clearly chimes with Bertrand Westphal's articulation of a geocritical approach to literary texts. According to Westphal, "geocriticism tends to favor a geocentered approach, which places *place* at the center of debate," and, as a result, "the spatial referent [either a named location or a generic form of topography such as the desert or the archipelago] is the basis for analysis, not the author and his or her work."[36] He also contends that it is "uncommon that artistic works are categorized according to the geographical spaces that they explore" and he suggests that "databases organized around spatial data are rare indeed." Westphal goes on to acknowledge that the "internet certainly helps" in the formation of "indices that associate a work with a place," but he also warns that "a great deal of patience, and a certain amount of scholarship, will be indispensable in forming a corpus necessary for a fully geocritical analysis."[37] Georeferencing Lake District landscape writing between 1750 and 1900, therefore, represents a first step toward the realization of this geocritical ambition. As a result, the literary thread to Spatial Humanities: Texts, GIS, Places moves away from what Westphal would define as the exclusively egocentric approach of Mapping the Lakes—which is structured around the mapping of subjective geographical accounts offered by just two canonical writers—to a "multifocalized" methodology which brings together dozens of interweaving, and frequently competing, examples of geospecific landscape writing.[38]

The analytic techniques described herein facilitate further thinking about how literary articulations of place have shaped the region's distinct cultural identity and have contributed to the sense of exceptionalism which has underpinned the concept of Lake District–ness since the middle of the eighteenth century. Alongside this, it allows users to identify a named location—such as the village of Rydal, where William Wordsworth lived from 1813 until his death in 1850—and to examine how the textual representations of that particular place evolved over a period of 150 years. By extension, the GIS-based spatial intertext opens up further geocritical thinking about the way in which writers, in representing place, respond to both the three-dimensional materiality of the spatial referent and (consciously or otherwise) earlier textual accounts of the named geographical site. The project then uses the tools described herein as a platform for exploring the role played by literary texts in the "placialization" of the Lakes: the term coined by Edward S. Casey to describe "the formation of place, for example, in landscape paintings and maps, but also in historical narration and prose fiction," and a term, therefore, which allows for the way in which literary texts inform a more general process of place-making.[39] Moreover, the use of layers of text—to use both the GIS and more general meaning of the term—allows analysis of the notion of the "stratigraphic," which is integral to Westphal's geocritical understanding of the literature of space, place, and landscape.[40] That is to say, by adding and removing multiple layers on the spatial intertext of the literary map, the user is able to examine imbrications of geography and temporality and to further his or her understanding of the "diachronic depths" of the region's geospecific literary history.[41]

The exploration of the collocation of place-names with specific semantic tags, described earlier, can, for instance, open up further thinking about the stratigraphic nature of Lake District landscape writing. It is possible, for example, to trace the textual origins, of the habitual practice of prefacing the Cumbrian fell-name, *Skiddaw* with the adjective *lofty* and to identify how successive generations of landscape writers reinscribe this collocation. Can the construction of the spatial intertext highlight any significant subversions of, and deviances from, this familiar descriptive tag? Can such changes be placed within the wider spatial history of shifting cultural attitudes to vertical topographies? As Svenja Adolphs points out, corpus approaches enable researchers to think further about two inextricably enlaced forms of intertextuality: the way in which texts allude to and echo previous writings, and, at the same time, the way in which the "semantic prosodies" of published texts relate to everyday language use.[42] When the "spatial referent" of the material landscape is introduced in order to form a triangulation of text, intertexts, and place, then it becomes clear that there is potential for exploring the relationship between Lake District landscape writing and vernacular geographies: the colloquial and quotidian articulations of place and spatial experience which often sit outside, and cut across, officially recorded geographies.[43] So, for instance, do Lake District landscape writers incorporate, within their texts, any locally generated names to describe the fell marked *Skiddaw* on the Ordnance Survey map? Do the topographical texts refer to an identifiably bounded geographical space when they name *Skiddaw*? Or, in using this toponym, do the texts allude to a vaguely conceived and imprecisely demarcated terrain? The use of GIS to explore the way in which vernacular geographies are embedded within literary texts can be expanded yet further by an examination of how, in linguistic terms, spatial practices are recorded and geographical directions are described. To return to the example of *Skiddaw,* how do writers record the physical movement toward this fell? Do particular expressions of spatial navigation become culturally entrenched within the spatial intertext? The use of the combination of corpus and GIS techniques, therefore, helps to move geocritically informed literary GIS research into the theoretical territory famously occupied by Michel de Certeau in his exploration of the roles played by spaces and places, maps and tours, in the formulation of the spatial stories which constitute the practice of everyday life.[44]

We are acutely sensitive, however, to the fact that these map-based approaches might appear to marginalize the practice of detailed textual analysis which was integral to the development of the literary GIS showcased in Mapping the Lakes.[45] In other words, it may appear as if the critical depth evidenced in the geographical readings, and mood mappings, of the texts by Gray and Coleridge is being sacrificed in favor of the surface spatial overviews with which the scholarly use of GIS has been traditionally associated but for which it has also been critiqued.[46] Spatial Humanities: Texts, GIS, Places, however, functions on multiple cartographic scales by oscillating between the mapping of overarching spatial patterns to the type of textual micromapping, which is more familiar to the literary critic. On the one hand, the project produces surface maps through the geovisualization of the evolution of linguistic patterns over time. As well as mapping and analyzing topographical texts by a diverse range of Lake District writers, the project simultaneously involves the spatialization of multiple versions of a single work of literary geography, such as William Wordsworth's prose *Guide to the Lakes* or Harriet Martineau's (critically neglected) *A Complete Guide to the English Lakes:* a process which illustrates how the multifocalization—which, for Westphal, can be understood by examining the work of a range of writers—can also be traced in the work of a single title which is revised and reconfigured over a period of time. The result is a multiscalar literary GIS which endeavors to allow for both the new reading practices opened up by the digitization and spatialization of large corpora and traditional attention to textual detail. Ultimately, then, Spatial Humanities: Texts, GIS, Places moves beyond Mapping the Lakes by developing a more fluid form of literary GIS which enables users to negotiate their own paths through the spatial intertext. The creation of the intertext will be founded upon a geocentric approach to literary history: the GIS has the potential to identify large-scale spatial patterns within Lake District literary geographies, and, at the same time, it visualizes, in cartographic form, the textual layering of particular named places. Yet, crucially, the system also enables users to pursue particular lines of inquiry by allowing for egocentric—or writer- and text-specific—points of entry.

The further advancement of the literary GIS can be traced by returning to an idea articulated in the introduction to this chapter: the possibility of combining the textual GIS with other multimedia representations of place in order to construct a deep map of the Lake District. According to David J. Bodenhamer, "deep mapping" offers "a fresh conceptualization of humanities GIS."[47] As Bodenhamer explains, "in its methods deep mapping conflates oral testimony, anthology, memoir, biography, images, natural history and everything you might ever want to say about a place, resulting in an eclectic work akin to eighteenth and early nineteenth-century gazetteers and travel accounts."[48] Many of the difficult-to-define Lake District landscape writings which we are digitizing and mapping neatly comply with Bodenhamer's definition of deep map, and our own layering of texts upon the geocentered literary GIS will lead to the further thickening of this sense of place. For Bodenhamer, though, authentically deep maps are predicated on several additional characteristics: "They are meant to be visual, time-based, and structurally open. They are genuinely multimedia and multilayered. They do not seek authority or objectivity but involve negotiation between insiders and outsiders, experts and contributors, over what is represented and seen."[49] A genuinely deep map of the Lake District, therefore, would necessarily involve the integration of a range of other geospecific materials including landscape paintings, historical maps, and oral histories. What is more, an authentically deep GIS would also need to provide opportunities for users to upload their own site-specific data and to contribute their own layers to the spatial palimpsest. Clearly, the creation of such an open and porous map presents challenges to the spatial humanities researcher. Do temporal and financial parameters make it possible to incorporate *all* of the spatial narrative forms and genres to which Bodenhamer refers? Is it necessary to introduce a control mechanism to monitor the quality of contributions to the site? By extension, is it essential to create two types of mapping to differentiate between the scholarly and the user-generated: a process which would clearly problematize the democratization of the GIS? Yet, in spite of such cautionary notes, there remains much that is attractive about this democratic form of mapping which can go some way to illustrating how the process of place-making is founded upon a complex interpenetration of material and imaginative, official, and vernacular geographies. The ongoing development of deep GIS, therefore, patently points the way toward new forms of practicing both digital humanities research and critical literary geography.[50]

CONCLUSION: TOWARD GEOGRAPHICAL TEXT ANALYSIS

We are approaching a situation where the georeferencing of very large quantities of text and, more importantly, their subsequent spatial analysis is becoming a possibility. These technical advances will bring with them some tensions as at least some of the approaches required to analyze very large datasets will be alien to current humanities research paradigms. Just because they are alien does not mean that they are wrong, however, and many lessons will have to be learned from the social sciences and elsewhere about their application. Human geography, which has a long tradition of both quantitative and qualitative approaches and has learned many lessons about the strengths and weaknesses of both, is particularly pertinent here.[51]

Briefly, however, there are implications at all three stages of geographical text analysis that this chapter has considered. The key question at the georeferencing stage is how accurate does this have to be? The process of moving from an automatically extracted list of suspected place-names to a layer in which every place-name that occurs in the text is linked to an accurate coordinate while all other names are discarded requires a level of manual intervention that is time-consuming and expensive—resources that are consequently being removed from the process of research and interpretation. Thus, rather than attempting to track down every error, the researcher is likely to be faced with a situation where the major errors that distort the pattern have been spotted—a process that is relatively simple when data are mapped, as this is a good way of noticing unlikely patterns—and the research phase can begin. During this phase, the researcher has to be aware that there will be errors in the data and that these will have to be updated as they are discovered. This may seem slightly uncomfortable; however, the social sciences are well used to dealing with situations where errors in databases have to be managed rather than eliminated. Equally the humanities are well used to thinking critically about their sources, and this approach merely extends this to the digital version.

In the analysis phase, corpus- and GIS-based techniques can both be used. Both sets of approaches have some tools that could be characterized as "broad but crude" and others that are "in-depth but narrow." The broad-but-crude techniques are those that attempt to quickly summarize the entire corpus, including techniques such as corpus-based frequency profiling or GIS-based density smoothing. They are broad in that they summarize the entire corpus, but crude in that they are highly abstracted from the original texts. The in-depth but narrow techniques include the use of concordances and map-based querying, which restructure the texts and present them in new ways. They are in-depth in that they still present the original texts to the researcher who has to develop his or her own understanding from them, and narrow in that using them is relatively slow and thus inevitably selective. Traditionally these approaches have been mutually antagonistic, but it should be clear that both have roles to play in understanding large quantities of georeferenced text. In particular, the broad techniques quickly summarize the corpus but are essentially descriptive. They can then point the researcher to where (in terms of within both the text and place) the in-depth techniques can be most usefully applied and these, in turn, provide more explanatory power. Referring back to the broader analyses then allows the researcher to ask where these lessons also apply, and where they appear to be different, thus contextualizing findings and avoiding the risk of atomistic fallacy, where a lesson learned from a small subset of data or places is inappropriately applied to the whole dataset or study area.

Finally, the digitization and spatialization of a relatively large corpus of landscape writing characterized by "transgeneric heterogeneity"[52] also highlights the integral role that GIS technology can play in the ongoing development of geocritical practice. The principal critical potentiality of GIS appears to be rooted in the facilitation of map-based reading of a corpus. That is to say, the large-scale mapping of significant quantities of texts can reveal abstract "shapes, relations, structures," to apply Franco Moretti's cardinal terms, which demand further research and explanation.[53] This process can lead to a reconfiguration of the corpus by drawing attention to previously marginalized or even neglected texts which, in actual fact, make a striking contribution to the spatial narrative of a particular geographical location such as the English Lake District, the whole of Britain, or potentially the whole world. Crucially, though, the exploratory methodologies set out in this chapter have been underpinned by the belief that the new reading practices, which are encouraged by the emergence of large-scale digital corpora, do not need to usurp traditional approaches to understanding texts. Instead, there is a need to think in terms of the new scales of reading which have been opened up by such digital corpora and to self-consciously reflect upon the ways in which the digital humanities researcher is able to move freely between the macro and the micro, the abstract and the concrete, the distant and the close.

ACKNOWLEDGMENTS

The research leading to these results has received funding from the European Research Council (ERC) under the European Union's Seventh Framework Programme (FP7/2007–2013)/ERC grant "Spatial Humanities: Texts, GIS and places" (agreement number 283850). Our thanks to Kirsten Hansen (Mt. Holyoke College, South Hadley, Massachusetts) for her work on the Wordsworth material used in figure 7.1.

NOTES

1. In reality, the situation is slightly more complex as approaches to the automated analysis of texts date back over thirty years; however, these are yet to be widely adopted, particularly within the humanities and social sciences.

2. Franco Moretti first articulated his model of "distant reading" in "Conjectures on World Literature," *New Left Review* 1 (2000), 54–68. For a useful summary of the critical controversy generated by Moretti's methodological proposal, see R. Serlen, "The Distant Future? Reading Franco Moretti," *Literature Compass* 7 (2010), 214–225.

3. There are many introductions to GIS available that describe this basic structure. See for example: N. R. Chrisman, *Exploring Geographic Information Systems,* 2nd ed. (Chichester, U.K.: John Wiley, 2002); K. C. Clarke, *Getting Started with Geographic Information Systems* (Upper Saddle River, N.J.: Prentice Hall, 1997); D. Martin, *Geographic Information Systems and Their Socio-economic Applications,* 2nd ed. (Hampshire, U.K.: Routledge, 1996).

4. I. N. Gregory, K. Kemp, and R. Mostern, "Geographical Information and Historical Research: Current Progress and Future Directions," *History and Computing* 13 (2003), 7–21.

5. See: I. N. Gregory and P. S. Ell, *Historical GIS: Technologies, Methodologies and Scholarship* (Cambridge: Cambridge University Press, 2007); or I. N. Gregory and R. G. Healey, "Historical GIS: Structuring, Mapping and Analysing Geographies of the Past," *Progress in Human Geography* 31 (2007), 638–653.

6. D. J. Bodenhamer, "The Potential of Spatial Humanities," in D. J. Bodenhamer, J. Corrigan, and T. M. Harris, eds., *Spatial Humanities: GIS and the Future of Humanities Scholarship* (Bloomington: Indiana University Press, 2010), 14–30.

7. Good introductions to this field include S. Adolphs, *Introducing Electronic Text Analysis: A Practical Guide for Language and Literary Studies* (London: Routledge, 2006); T. McEnery and A. Hardie, *Corpus Linguistics: Method, Theory and Practice* (Cambridge: Cambridge University Press, 2012); and C. Manning and H. Schütze, *Foundations of Statistical Natural Language Processing* (Cambridge, Mass.: The MIT Press, 1999).

8. See: I. N. Gregory and A. Hardie, "Visual GISting: Bringing together Corpus Linguistics and Geographical Information Systems," *Literary and Linguistic Computing* 26 (2011), 297–314; I. N. Gregory and D. Cooper, "Thomas Gray, Samuel Taylor Coleridge and Geographical Information Systems: A Literary GIS of Two Lake District Tours," *International Journal of Humanities and Arts Computing* 3 (2009), 61–84; D. Cooper and I. N. Gregory, "Mapping the English Lake District: A Literary GIS," *Transactions of the Institute of British Geographers* 36 (2011), 89–108.

9. Spatial Humanities: Texts, GIS, Places project is available at http://www.lancs.ac.uk/spatialhum (accessed August 14, 2013).

10. The account of corpus linguistics given in this section is drawn largely from McEnery and Hardie, *Corpus Linguistics*.

11. For example, the IMPACT Project (http://www.impact-project.eu) and Early Modern OCR Project—eMOP (http://emop.tamu.edu) as well as the Text Creation Partnership (http://www.textcreationpartnership.org) (all accessed August 14, 2013).

12. The Lancaster Newsbooks Corpus is available at http://www.lancs.ac.uk/fass/projects/newsbooks (accessed August 14, 2013).

13. See R. Garside, G. Leech, and T. McEnery, eds., *Corpus Annotation: Linguistic Information from Computer Text Corpora* (London: Longman, 1997).

14. L. L. Hill, *Georeferencing: The Geographic Associations of Information* (Cambridge, Mass.: The MIT Press, 2006) provides a comprehensive overview of the use of gazetteers especially in chapter 5.

15. Geonames is available at http://www.geonames.org (accessed August 14, 2013).

16. World-Gazetteer is available at http://www.world-gazetteer.com (accessed February 21, 2012).

17. Getty Thesaurus of Geographical Names is available at http://www.getty.edu/research/tools/vocabularies/tgn/index.html (accessed August 14, 2013).

18. Ordnance Survey's 1:50,000 Gazetteer is available at http://www.edina.ac.uk/digimap (accessed August 14, 2013).

19. A layer is effectively the GIS equivalent of a database table. The difference is that in a layer every row of data in the table (termed *attribute data*) is linked to a map-based location which is underlain by coordinates (termed the *spatial data*).

20. C. Grover, R. Tobin, M. Woollard, J. Reid, S. Dunn, and J. Ball, "Use of the Edinburgh Geoparser for Georeferencing Digitized Historical Collections," *Philosophical Transactions of the Royal Society A* 368 (2010), 3875–3889 provides an example of this.

21. See C. Gabrielatos and P. Baker, "Fleeing, Sneaking, Flooding: A Corpus Analysis of Discursive Constructions of Refugees and Asylum Seekers in the UK Press (1996–2005)," *Journal of English Linguistics* 36 (2008), 5–38.

22. P. Rayson, D. Archer, S. L. Piao, and T. McEnery, "The UCREL Semantic Analysis System," *Proceedings of the Workshop on "Beyond Named Entity Recognition Semantic Labelling for NLP Tasks" in Association with 4th International Conference on Language Resources and Evaluation (LREC)* (Lisbon: LREC, 2004), 7–12.

23. Ibid, 7.

24. See http://www.lancs.ac.uk/mappingthelakes/v2 (accessed August 14, 2013) for a draft of the use of Google Maps; http://www.lancs.ac.uk/mappingthelakes/Interactive%20Maps%20Introduction.html (accessed August 14, 2013) provides an earlier example of how Google Earth can be used in the same way.

25. D. J. Cohen and R. Rosenzweig, *Digital History: A Guide to Gathering, Preserving, and Presenting the Past on the Web* (Philadelphia: University of Pennsylvania Press, 2006).

26. T. C. Bailey and A. C. Gatrell, *Interactive Spatial Data Analysis* (Harlow, U.K.: Longman, 1995).

27. Gregory and Hardie, "Visual GISting: 297–314."

28. See: Gregory and Hardie, "Visual GISting," 297–314 and Cooper and Gregory, "Mapping the English Lake District," 89–108.

29. Good texts on point pattern analysis include: Bailey and Gatrell, *Interactive Spatial Data Analysis*; A. S. Fotheringham, C. Brunsdon, and M. E. Charlton, *Quantitative Geography: Perspectives on Spatial Data Analysis* (London: Sage, 2000); and C. Lloyd, *Spatial Data Analysis: An Introduction for GIS Users* (Oxford: Oxford University Press, 2010).

30. L. Anselin, "Local Indicators of Spatial Association—LISA," *Geographical Analysis* 27 (1995), 93–115.

31. A. S. Fotheringham, C. Brunsdon, and M. E. Charlton, *Geographically Weighted Regression: The Analysis of Spatially Varying Relationships* (Chichester, U.K.: Wiley, 2002).

32. See http://www.lancaster.ac.uk/mappingthelakes (accessed August 14, 2013).

33. Cooper and Gregory, "Mapping the English Lake District," 89–108; and Gregory and Cooper, "Thomas Gray, Samuel Taylor Coleridge and Geographical Information Systems," 61–84.

34. For more on sentiment analysis research see, for example, S. Piao, Y. Tsuruoka, and S. Ananiadou, "Sentiment Analysis with Knowledge Resource and NLP Tools," *International Journal of Interdisciplinary Social Sciences* 4 (2009), 17–28.

35. The conceptualization of the archive as "a textual environment [which is] open to continuous transformation and development" owes a theoretical indebtedness to Jerome McGann's early exploratory work on the Rossetti Archive: J. McGann, *Radiant Textuality: Literature after the World Wide Web* (New York: Palgrave, 2001), 82.

36. B. Westphal, *Geocriticism: Real and Fictional Spaces*, trans. Robert T. Tally Jr. (New York: Palgrave Macmillan, 2011), 112–113.

37. Westphal, *Geocriticism*, 117.

38. Ibid., 122.

39. E. S. Casey, *Representing Place: Landscape Painting and Maps* (Minneapolis: University of Minnesota Press, 2002), 351.

40. Westphal, *Geocriticism*, 137.

41. Ibid., 139.

42. Adolphs, *Introducing Electronic Text Analysis*, 68–69.

43. See, for example, A. J. Evans and T. Waters, "Mapping Vernacular Geography: Web-based GIS Tools for Capturing 'Fuzzy' or 'Vague' Entities," *International Journal of Technology, Policy and Management* 7 (2007), 134–150. See also ongoing collaborative research by the Ordnance Survey, http://www.ordnancesurvey.co.uk/education-research/research/vernacular-geography.html (accessed August 14, 2013).

44. M. de Certeau, *The Practice of Everyday Life*, trans. Steven Rendall (Berkeley: University of California Press, 1984), 115–130.

45. F. Moretti, *Graphs, Maps, Trees: Abstract Models for a Literary History* (London: Verso, 2005), 1.

46. For and overview see: J. Pickles, "Arguments, Debates, and Dialogues: The GIS-Social Theory Debate and the Concern for Alternatives," in P. A. Longley, M. F. Goodchild, D. J. Maguire and D. W. Rhind, eds., *Geographical Information Systems: Principals, Techniques, Management and Applications*, 2nd ed. (Chichester, U.K.: John Wiley, 1999), 49–60.

47. D. J. Bodenhamer, "The Potential of Spatial Humanities," 27.

48. Ibid., 27.

49. Ibid., 27.

50. A. Thacker, "The Idea of a Critical Literary Geography," *New Formations* 3 (2005–2006), 56–73.

51. A number of good general readers on human geography are relevant here including: D. N. Livingstone, *The Geographical Tradition: Episodes in the History of a Contested Enterprise* (Oxford: Blackwell, 1992); R. J. Johnston, *Philosophy and Human Geography: An Introduction of Contemporary Approaches* (London: Edward Arnold, 1983); A. Holt-Jensen, *Geography: History and Concepts: A Student's Guide*, 3rd ed. (London: Sage, 1999).

52. Ibid., 119.

53. Moretti, *Graphs, Maps, Trees*, 1.

8

GIS AS A NARRATIVE GENERATION PLATFORM

MAY YUAN, JOHN MCINTOSH,
AND GRANT DELOZIER

INTRODUCTION

Maps have long been one of the key tools to represent the landscape within which histories occurred. While being static, maps present the spatial dimension of historical data and reveal spatial associations among spatial features of interest. Much research in spatial histories or historical geographical information systems (GIS) rises to the challenge of visualizing historical social data, geocoding historical cultural landmarks, and analyzing their spatial patterns over time.[1] Yet, historical investigations go far beyond thematic or statistical mapping. Historian John Gaddis noted that historians exercise *selectivity, simultaneity,* and *shifting of scale* in manipulation of space and time to construct narratives that interpret the past.[2] Selectivity is necessary so that historians can simplify a complex reality into something manageable for a study. When selected events expand over space and time, historians examine multiple places at once (i.e., simultaneity) and shift scales when they use a particular episode to make a general point. *Scale shifting* is a fundamental tool for narration in history, and *simultaneity* leads to the study of histories as mapping the past landscape. Landscape patterns of historical events constitute the structure that historians observe in the present. Narratives are being developed as historians interpret the processes that produced the landscape structure. To Gaddis, historians embed generalizations in narratives, while social scientists embed narratives in generalizations. Instead of categorical causes, historians emphasize contingent causes that are responsible for developing singularities in continuity and lead to particular generalizations in history.

This study on GIS as a narrative generation platform (i.e., narrative GIS) aligns well with Gaddis's landscape of history. Narratives are sequential organizations of events. While conventional GIS centers on information characterizing geographies, a narrative GIS aims to represent and order events in support of constructing spatial narratives. Here, spatial narratives refer to meaningful sequences of spatial events. Our premise is that locations where events took place should be considered equally important to the time of occurrences in history, and the treatment of time should be considered as important as locations in geography. A narrative GIS henceforth aims to provide the necessary framework to make connections among events in space and time for narrative generation. GIS supports the need for selectivity in historical studies by database queries to retrieve events of interest, and for simultaneity, by mapping selected events across space and time and contextualizing these events with geographic features or other events. GIS zoom functions enable historians to shift across local, regional, and global scales insofar as the data permit. The vision of a narrative GIS is to facilitate the interactive selection of events of different types, relate events across space and time, and assess microscopic and macroscopic structures to embed multiple generalizations in narratives that help explain the underlying historical processes.

A narrative is a meaningful sequence of events, and therefore events are basic constructs for narrative generation. Consequently, event objects are essential to a narrative GIS database. While there is no universal definition of events, we simply consider spatial event objects as a quadruple of [*actor, action, location,* and *time*].[3] *Actors* are entities involved in an event, and *actions* evoke happenings of an event at a location and time. Actors may be biotic (e.g., humans or animals), abiotic (e.g., fires or rocks), or immaterial (e.g., communications or ideas). Actors take actions, and actions drive changes to properties, conditions, or locations. Events are innately temporal; event history modeling, for example, analyzes and projects the timing and periodicity of event occurrences.[4] While most events are spatial, spatial markers may not be explicit and are treated as add-ons in most event studies.

Nevertheless, many scholars recognize the importance of spatial dimensions to event modeling. For example, the duration of civil wars in various countries is commonly modeled with statistical regressions on the magnitude and frequency of conflicts; yet additional spatial considerations led to new insights that "the greater the frequency of states bordering the civil war state, the longer the duration of the civil war."[5] In other words, the number of bordering states has a prolonging effect on the duration of the civil war in a state. With equal treatments of space and time in representing and analyzing events, a narrative GIS is expected to provide new insights into the correlation, interaction, and structure of historical events and narratives in space and time. Narrative generation is performed by connecting events in space and time based on actors, actions, or both to decipher the spatiotemporal relationships among actors and actions in making histories. As Gaddis noted, historians embed multiple generalizations in narratives. Likewise, a narrative GIS generalizes spatial events in the form of quadruples [actor, action, location, time] and embeds multiple generalizations of ordered spatial events to generate narratives.

Since the majority of historical data are documents, our development of a narrative GIS begins with ingesting text documents, extracting and assembling spatial events to form GIS event databases, querying and structuring events to generate spatial narratives, and storing spatial narratives for future queries and analyses (figure 8.1). As proof of concept, we use two distinctive corpora of histories in building narrative GIS databases and narrative analytics: Dyer's *Compendium of the War of the Rebellion* and the *Richmond Daily Dispatch*. Frederick H. Dyers, a Civil War veteran compiled the *Compendium* based on materials from the Official Records of the Union and Confederate Armies and other sources.

Dyer's Compendium[6] lists organizations and movements of regiment cavalries mustered by state and federal governments for services in the Union armies. The second source of historical documents is the *Richmond Daily Dispatch* provided by the Digital Scholarship Laboratory at the University of Richmond. The newspaper was one of the most widely distributed newspapers in the south during the Civil War and included news from the entire east coast. The *Richmond Daily Dispatch* retained the reputation of being politically unbiased and was published throughout the Civil War. The use of the two document sources serves two pur-

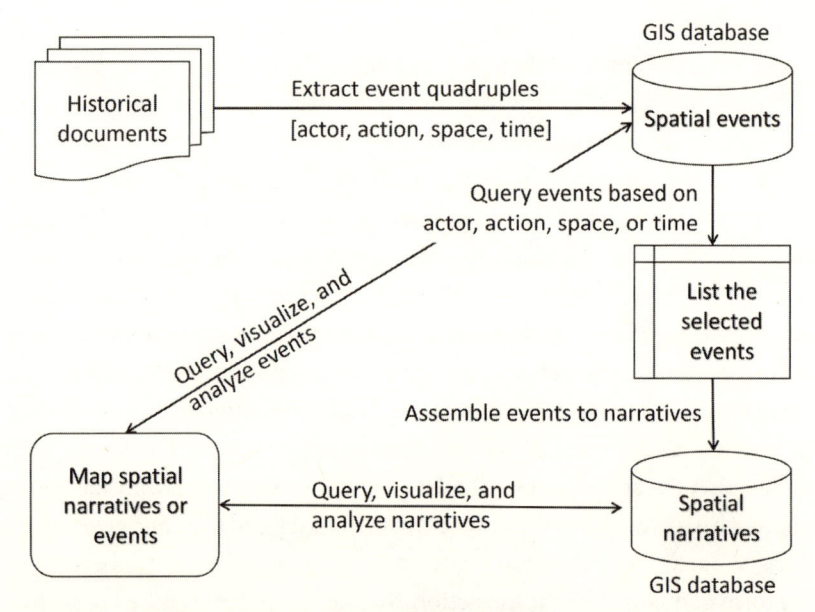

FIGURE 8.1. Workflow for narrative GIS development.

poses. First, the writing styles are distinctive, and therefore, provide the challenge of developing algorithms of text analytics that are not specific to particular source documents. *Dyer's Compendium* has a very concise writing style, and since actors in descriptions are declared in respective titles, actors are seldom noted explicitly in sentences. Consequently, most sentences in *Dyer's Compendium* have no subjects, and sometimes, action nouns (such as *movement*) are used in place of action verbs (such as *move*). All sentences in one document of *Dyer's Compendium* refer to the same military unit, such as Alabama First Regiment Calvary, and the actor is not repeated in the article.

The *Richmond Daily Dispatch*,[7] on the other hand, contains news reports, public announcements, and advertisements, and the writing styles and use of words in the Civil War era often do not conform to the contemporary conventions. Text analytics rely upon a corpus to develop and test algorithm performance. Similar to sample data in science, a corpus is a large set of texts to serve as the basis for statistical analysis and hypothesis testing in text analytics. Sample data shall be randomly drawn from the population of documents so that the statistical characteristics of the sample are representative of the population. Likewise, the scope and content of a corpus can limit the generalizability of the text analytics algorithms developed based on the corpus. Use of the two distinctive documents is challenging for text analytics here because the existing corpora in various packages of text analytics have been developed as gold standards for contemporary texts. Besides the algorithmic challenge, the second purpose to include the two types of documents is to demonstrate the fusion of events from different sources for narrative generation in a GIS. The prospectus of correlating events from different sources provides opportunities for spatiotemporal mashups of histories that are cognitively demanding in manual operations.

DATA USED IN THE NARRATIVE GIS DESIGN

As illustrated in figure 8.1, the first step in building a narrative GIS is to extract event quadruples from historical sources, including both structured and unstructured data. Many national historical GIS projects provide rich diverse historical data for a nation, such as the U.S. Historical GIS[8] the Great Britain Historical GIS,[9] the Netherland Historical GIS,[10] and the China Historical GIS.[11] The Electronic Cultural Atlas Initiative,[12] Arts and Humanities Data Service,[13] Council of European Social Science Data Archives,[14] and many other regional or international organizations have substantially contributed to structured historical data. Yet, massive amounts of historical sources, including official documents, newspapers, journals, and many text materials, are unstructured data, which a narrative GIS should be able to ingest.

Both *Dyer's Compendium of the War of the Rebellion* and the *Richmond Daily Dispatch* are unstructured data, and their text styles are quite distinct. *Dyer's Compendium* concisely lists regiment movements and battles in every state from April 12, 1861, to May 6, 1866. The *Compendium,* for example, details the start of the Alabama First Regiment Cavalry at Montgomery in November 1861, the battles in which the cavalry engaged as well as the movements and experiences of the cavalry. Our study includes a total of 3,430 files from *Dyer's Compendium* in the raw data. Each file corresponds to an infantry, a cavalry, or artillery of union regiments.

On the other hand, the *Richmond Daily Dispatch* articles provided updates of events in Richmond, the state of Virginia, and the Confederacy at large, news about movers and shakers of the Confederate government and military, battles in Montgomery and New Orleans, the "mobocracy" in New York City, events in Europe, even local accidents or severe weather in other states, or official announcements/notices and commercial advertisements during the Civil War. Our study incorporates a total of 1,385 *Daily Dispatch* files published from November 1, 1860, to December 30, 1865, including those missing dates.

METHODS TO EXTRACT EVENTS

Our assumption is that an action denotes an event. Extraction of spatial events in texts starts with identification of action verbs (such as *move, bring,* or *escape*) or action nouns (such as *engagement, construction,* or *retreat*). In addition, the other three elements of an event quadruple: actor, location, and time, need to be determined in association with the identified action verb or action noun. Hence, our workflow of event extraction consists of six key steps: (1) determine text analysis units; (2) identify action verbs or action nouns; (3) identify time references and text units; (4) identify location references and text units; (5) combine all identified elements into a GIS database; and (6) build spatial and temporal relationships among events and narrative objects. Figure 8.2 provides an overview of the workflow for text analytics and event extraction to assemble a geodatabase with events and narratives.

Input texts are categorized into units of self-contained articles (i.e., text analysis units) in which information follows a consistent topic. Examples of a text analysis unit include a chapter or a news report within which the messages among actors, actions, location references, and time references remain coherent. For each text analysis unit, we applied the Natural Language Toolkit (NLTK)[15] to tokenize sentences and identify parts of speech (e.g., nouns, verbs).

The parts of speech as well as their order of appearance in a sentence serve as the basis for extracting event quadruples. The first step is to identify action verbs indicative of event occurrences. Sentences with modal verbs and stative verbs are removed from further consideration. Follow-up procedures relate spatial and temporal markers in the respective text analysis units. Some sentences may include locations and times explicitly with the action verbs. Other sentences without direct space and time indicators

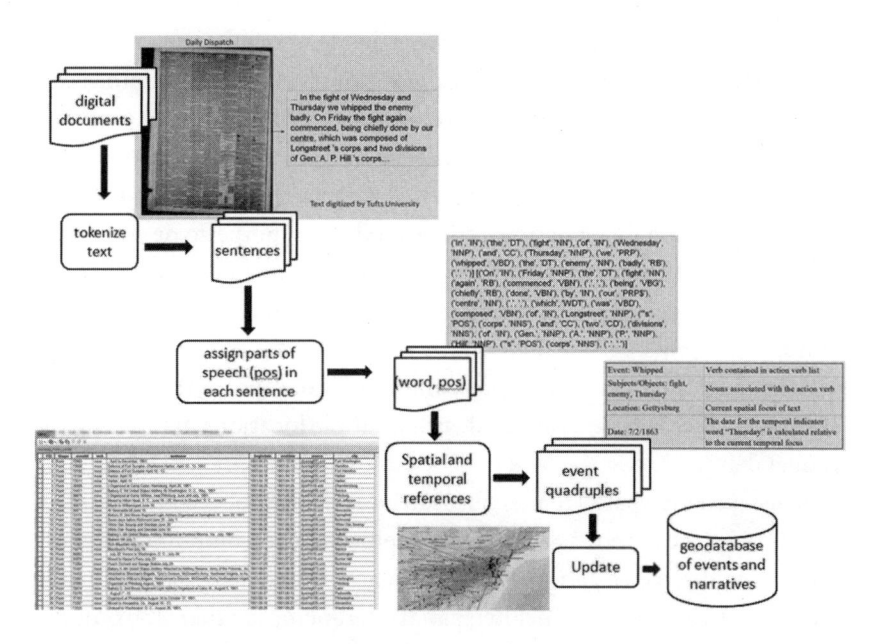

FIGURE 8.2. Workflow to tokenize text, assign parts of speech, and assemble events for a geodatabase.

require inferences from texts ahead. In other words, we assume that events took place at the same locations unless noted otherwise. Moreover, there may be no or more than one actor (i.e., nouns indicative of subjects or objects) referenced in a sentence. To avoid losing context, we include the original sentence and the article identifier with each event quadruple in a geodatabase so that the user can always reference to the text for further interpretation. Event quadruples are considered as atomic events since they represent the primitives of events denoted by individual action verbs. Users can select events of interest to assemble event groups or sequence events for narrative generation. Atomic events, event groups (such as a war of many battles), and narratives are stored in hierarchies in a geodatabase.

DETERMINE SPATIAL REFERENCE

We use *spatial reference* as a general term that means to reference something spatially. There are many related terms to spatial reference, such as geoparsing, geotagging, toponym recognition, toponym resolution, geocoding, georeferencing, and potentially other terms in the literature. Each of these terms emphasizes different aspects of spatial reference processes, and our procedures mostly relate to toponym recognition and toponym resolution. As we assume that each action verb (or action noun) indicates an atomic event, spatial referencing is to determine locational information associated with a given action verb, excluding hypothetic or suggestive sentences in which actions may not have taken place and therefore events may not have occurred. Some action verbs or action nouns account for multiple locations. For example, move or movement is often with locations of origin, destination, and perhaps intermediate stops. In these cases, an event identifier will be associated with multiple locations in building the geodatabase.

Conceptually, spatial referencing can be a simple process: identify place names (or place markers, such as geographic features, monuments, or addresses) and retrieve geographic coordinates of the place as specified in known sources. Algorithmically, spatial referencing is a two-step process: (1) identify place names in a text, and (2) determine the coordinates of the place (for example, looking up the place name in gazetteers to retrieve geographic coordinates or calculate the coordinates based on a reference system such as addresses). However, spatial referencing inherits numerous challenges from two primary sources of complexity about place names: (1) non-geo ambiguity: place names may be confused with other names or place names may be used as metonyms in a text, and (2) geo ambiguity: multiple places may have the same name in gazetteers.

IDENTIFY PLACE NAMES IN TEXT

The complexity of place names manifests itself semantically, geographically, and historically. Place names can be difficult to differentiate from person names, title names, feature names, organization names, and other proper nouns in a text. Identification of historical place names encounters yet additional challenges.[16] Locations may change names or change spellings over time; settlement places may change locations over time, and hence a place name may be referenced to different locations over time; and places may be relative to landmarks that no longer exist or cannot be located. Furthermore, procedures need to distinguish between events that are stationary and events that involve multiple places. An example of a stationary event is that President Abraham Lincoln delivered a speech at Gettysburg, Pennsylvania, November 19, 1863. The New York Ninth Regiment Cavalry moving from Hunterstown to Gettysburg on July 2, 1863, is an event of multiple locations and hence should be mapped as a sequence of vectors connecting locations in a temporal order.

Since all place names in English are proper nouns, we have applied the NLTK to assign parts of speech to each word in a sentence for all sentences in the input documents (figure 8.2).[17] We have developed several tests to identify place names. The first test checks proper nouns with a list of common people and entity names.[18] Proper nouns not on the list of common names for people or entities (e.g., organization or title) proceed to the next test that checks the presence of spatial propositions, apostrophes, and determiners. Spatial propositions are common precursors to place names in English, so the successor of a spatial proposition is a place name candidate. Proper nouns followed by spatial prepositions are assumed to be place names, but those with by apostrophes or determiners are not. Proper nouns are also tested against a list of state names or state abbreviations. These proper nouns which surround a place name are used as contextual words to narrow the geographic scope of the place. If a contextual word pertains to a state name, for example, the place name nearby in a sentence is then assumed to fall within the state.

Another approach to identify place names in text is named entity recognition (NER)[19] (aka named entity classification) that sorts names for persons, organizations, locations, and other entities in texts through statistical regressions or data mining methods. The performance of a NER algorithm by and large depends upon the size and richness of an annotated corpus tailored to the nature of the texts from which named entities are to be extracted. The annotated corpus is commonly referred to as the gold standard for NER training to specify atomic elements in the texts of interest, commonly limited to a particular genre or domain. Applications of a NER tool to a new genre or domain can decrease the performance by 20 to 40 percent[20] in precision and recall.[21] Construction of a gold standard corpus is a laborious task, and often a customized gold standard is necessary to ensure a good NER performance. We developed a NER classifier based on a novel set of linguistic features and naïve Bayes methods to identify place names in texts. We choose the naïve Bayes classification method for its performance and its differentiation of false-positive and false-negative errors. False-positive errors can be subsequently eliminated through gazetteer matching. We devised a set of novel features to note contextual and semantic information in sentences from 112 randomly selected *Daily Dispatch* articles. The feature set was composed of a two-by-two parts of speech window, the named entity itself, a spatial phrase test, the dominant semantic domains of nearby verbs, and a spatial words test. The naïve Bayes classifier examined the combinatory features to determine if the named entity was a place name.

Our preliminary test resulted in 85 to 90 percent accuracy on place name identification in the *Daily Dispatch*. Since *Dyer's Compendium* has a distinctively concise writing style and commonly contains incomplete sentences, the NER classifiers developed using *Daily Dispatch* texts will perform inadequately in identifying place names from *Dyer's Compendium*.

DISAMBIGUATE PLACE NAMES

The multiplicity of place names poses another layer of spatial referencing challenges. Many locations may have the same name, some place names are only used locally or regionally (e.g., the corner stone), and some place names are in reference to landmark features, such as posts, fence lines, trees, rivers, and so on. Spatial hierarchies and multiple levels of spatial references, such as countries, cities, villages, and such, lead to uncertainty in spatial representation of place names,[22] when multiple geographic units across scales share the same place name (such as Cleveland is both a city name and a county name).

Digital gazetteers and geospatial datasets provide a wide range of place names to match with the identified proper nouns hierarchically in the order of state, county, city, and others if they exist. Specifically, we included the following sources for place name matching: (1) U.S. populated places, U.S. Census Bureau's hundred most-populated cities by decade, building, locale, military, and valley portions in Geographic Names Information System; (2) U.S. rivers and streams edited/annotated from U.S. Geological Survey Hydrography; (3) historical U.S. states, territories, and counties from National Historical GIS; (4) world administrative boundaries at the province/state level and U.S. lakes edited/annotated from Natural Earth; and (5) continents, world countries as of 2010, and highly populous world cities from the digital chart of the world by Esri. In addition, we created two datasets to address special needs for the documents used in the project: (1) U.S. Civil War battlefields crawled from Wikipedia, and (2) a regions file representing regions of the United States during the Civil War.

For proper nouns with multiple matches (e.g., *Georgetown* is shared by over 70 locations among U.S. cities), we have developed three rules of thumb: (1) spatial proximity advantage (give preference to the closest place from the previous geocoded place in the text), (2) population dominance (give preference to the cities with larger populations), and (3) context advantage (give preference to the city within the state name, county name, or other place name found in the contextual words. When a place name has multiple matches in gazetteers, the spatial proximity rule assumes that the city geographically closer to the previously identified location in the text is more likely to be the referenced city than those farther away. For example, if the previous text is referenced to St. Louis, then the place name *Miami* will be referenced to Miami, Oklahoma, instead of Miami, Florida. The rule of population dominance is implemented by a list of the hundred most-populated cities during the Civil War period. Spatial dominance rule gives priorities to cities in the U.S. populated places gazetteers under the assumption that cities with larger populations are more likely to be noted without any previous spatial references. Hence, when there is no geographic indication, the proper noun *Cleveland* is more likely to be Cleveland, Ohio (2010 census population: 396,815) than Cleveland, Tennessee (2010 census population: 41,285).

All the three rules of thumb are combined into one derived *comparison distance* to determine collective effects:

Comparison distance = Euclidean distance × (1-context_weight) × (1-population_weight)

In the equation, *Euclidean distance* is the straight line distance between the city of consideration and the previous coded city. In the example of Miami, it will include the Euclidean distance from St. Louis to Miami, Oklahoma, and the Euclidean distance from St. Louis to Miami, Florida. *Context weight* accounts for the contextual advantage of other places surrounding the place name in question in the same sentence (e.g., if a sentence includes Oklahoma). *Population weight* relates to whether cities are in the list of the hundred most-populated cities (e.g., Miami, Florida is, but Miami, Oklahoma is not). Higher context weight or population weight will result in a greater reduction of Euclidean distance and hence smaller comparison distance. The place name in question is spatially referenced to the city with the smallest comparison distance of all other cities with the same place name. Assignments of weights can be empirically based or subjectively determined. In our study, preliminary tests suggest that a higher population weight (e.g., 90 percent) works better for texts from the *Daily Dispatch* but a lower population weight (e.g., 50 percent) performs better on texts from *Dyer's Compendium*. More systematic research is needed to determine the optimal values for weight assignments.

The three rules of thumb for place disambiguation help narrow in the most probable place name match among locations with the same place name. There are additional rules and strategies to enhance the precision and recall of spatial referencing, and many studies, including ours, are seeking ways to make significant improvements.

DETERMINE TEMPORAL REFERENCE

Temporal referencing determines time markers for events, including dates, months, and years. In addition to occurrence and duration, temporal reference information facilitates event ordering for narrative generation. Temporal referencing can be based on *explicit time markers* or *deictic time markers*. Explicit time markers are chronological times explicitly noted as absolute clock time, dates, months, years in the Gregorian or other calendar systems. Deictic time markers are context-dependent, such as yesterday, last night, or three years ago, and deictic markers must rely on identifying both the proper explicit time marker to anchor the temporal reference and the associated temporal relationships with the explicit marker to determine temporal reference.

We have developed the following procedures to extract the anchored time markers and infer time references for deictic time markers in determining temporal references for events. Anchored time markers can be explicitly noted in the document (e.g., *Dyer's Compendium*) or the date of publication for a newspaper (e.g., *Daily Dispatch*). Once an explicit time marker is found, the marker will serve as an anchored time marker to relate all subsequent events, unless a new explicit time marker is identified. The new explicit time marker is then the anchored time marker. When no explicit time marker can be used for reference, events are ordered relatively according to thirteen temporal relationships defined by James Allen[23] whose model has been broadly applied in temporal information systems and temporal reasoning and analysis. When explicit time markers are available, temporal relationships are established based on their chronological order and duration measures. Otherwise, time-relevant prepositions (e.g., before or afternoon) or adverbs (e.g., faster or earlier) are used for temporal ordering. Our preliminary test suggests that our temporal referencing program performs reasonably well: of 1,698 input articles, all anchor time data are correctly identified and six events received no time references.

USE CASE EXAMPLE

A narrative GIS test bed was developed to experiment with the design of spatial narrative generation and functions for text analytics, spatial referencing, temporal referencing, event search, and mapping. The following example goes through procedures that demonstrate the current stage of GIS for spatial narrative generation. While simplistic, the test bed shows promise for expanding current GIS technology to a spatial narrative generation platform.

The test bed was built with Python programming language with links to tools and functions in NLTK and WordNet.[24] WordNet relates user input of action verbs or nouns to synonyms so that action verbs of similar concepts can be returned for user selection. The use case explores events in which slaves ran away and were announced in public announcements or advertisements posted by slave owners for cash rewards in the *Daily Dispatch*. Using the attribute filters tool (figure 8.3) the search initiates with the verb *run* and the noun *negro*.[25] The system retrieves synonyms from WordNet, and the user can determine which verbs or nouns to be used for event search. The user also selects the articles for the search. In this case, no specific data source is selected which indicates that the search will go through all data sources from the *Daily Dispatch* and *Dyer's Compendium* in the study.

The interface contains two additional tools to refine queries: range filters and relationship filters (figures 8.4). The range filters help specify the state or period of interest, so that only events within the specified states or

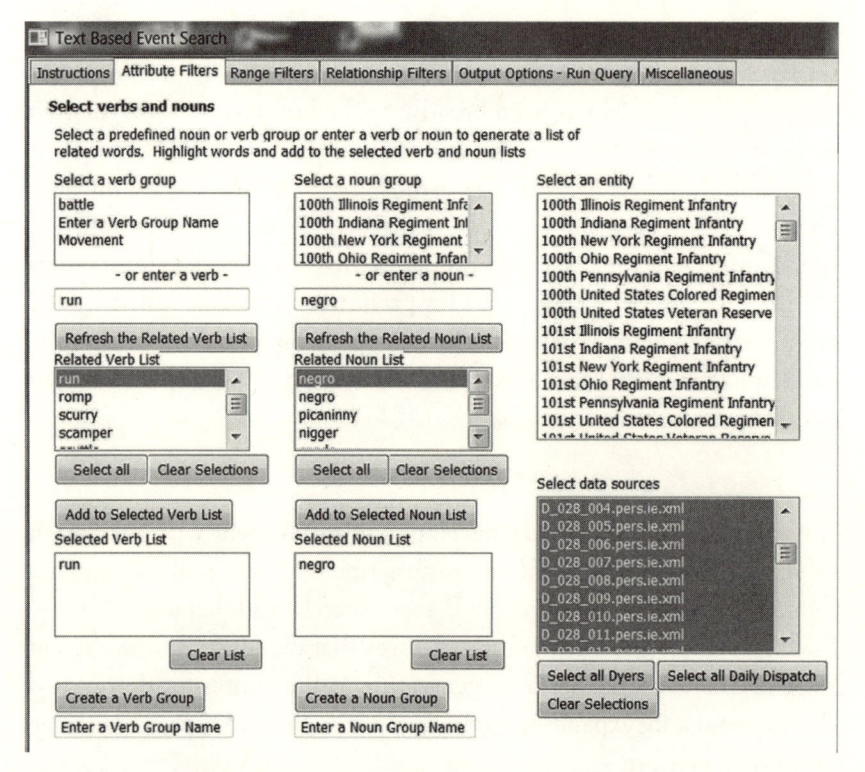

FIGURE 8.3. Attribute filters interface for the narrative GIS test bed. The user first specifies the verb and noun of interest. "Related verb list" and "related noun list" are synonyms retrieved from WordNet. The user also specifies the articles to search for the event of interest.

during the specified period of time will be returned to the user. With the relationship filters, the user can set spatial limits to search for events, such as within a specified distance of selected cities or locations and a specified time of interest. If there are prestored spatial narratives, the search can also be based on the spatial and temporal extents of the spatial narratives of interest. If none are specified, the search will include all events based on specifications in the attribute filters.

Figure 8.5 shows the results from the case study. There are 221 records retrieved from 1,384 articles from the *Daily Dispatch* and 3,431 articles from *Dyer's Compendium*. Among the 221 atomic events, 41 events cannot be spatially referenced. Starting dates for the 221 events were with temporal

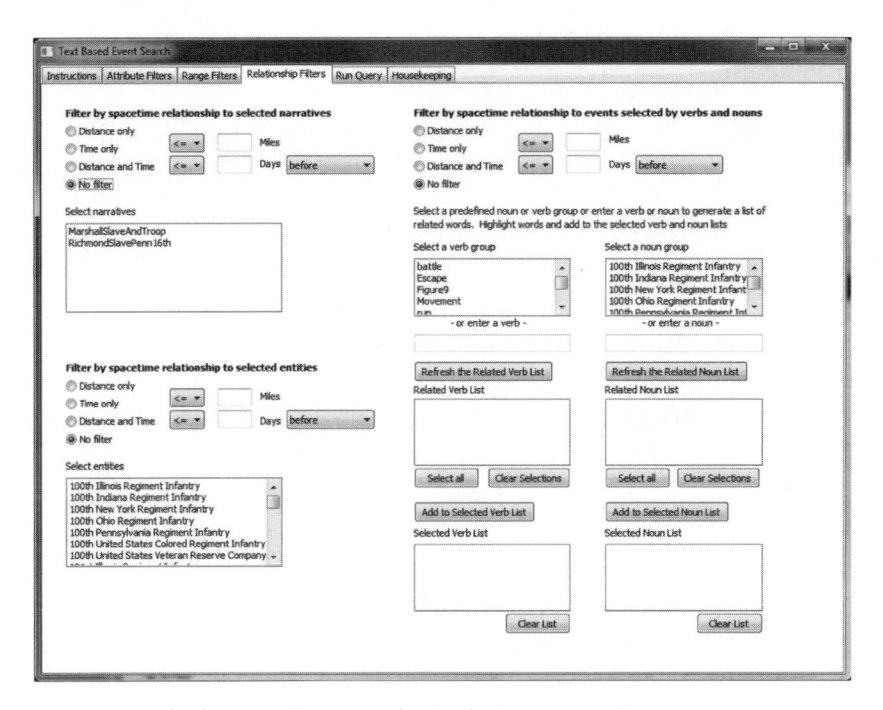

FIGURE 8.4. (*top*) Range filters and (*bottom*) relationship filters to refine an event search.

references from July 11, 1852, to December 31, 1964. While the search went through both data sources, all the retrieved events are originated from the *Daily Dispatch* articles. *Dyer's Compendium* details the movement of the Union armies and reasonably excludes other types of events. On a yearly basis, the year of 1861 marks a big increase in reporting of slave-run events,

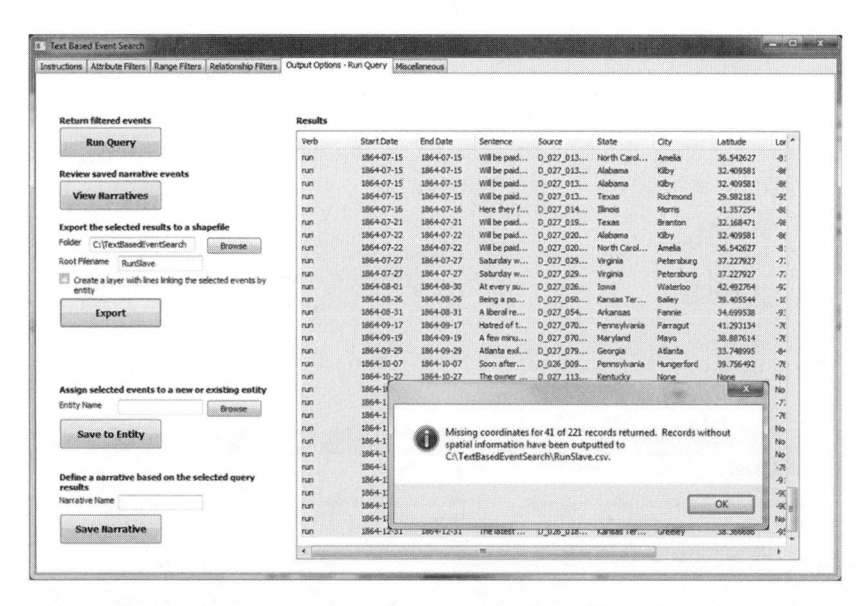

FIGURE 8.5. Atomic events on slave run retrieved from all articles included in the study.

and before that only two such events were reported in 1852 (figure 8.6A). The year of 1861 marked the start of the Civil War: Abraham Lincoln was newly elected to the presidency and South Carolina left the Union in January; the Confederate States of America formed in February; Lincoln was inaugurated in March; West Virginia formed in June to defy the Union; and President Lincoln revokes Gen. John C. Frémont's unauthorized military proclamation of emancipation in Missouri in September; and three battles (Leesburg, Ball's Bluff, and Harrison) were fought in October. No claim is made here to any connection between these major developments in the Civil War and runaway slaves. The case is only to demonstrate that slave-run events or other social events extracted from historical documents may hint at correlation in space and time which furthermore may lead to new historical insights.

Preliminary spatial referencing suggests locations of these slave-run events. The narrative GIS provides an export function to create a shapefile of selected events with geographic coordinates from matched places in digital gazetteers (figure 8.5). Figure 8.7 shows the event locations as

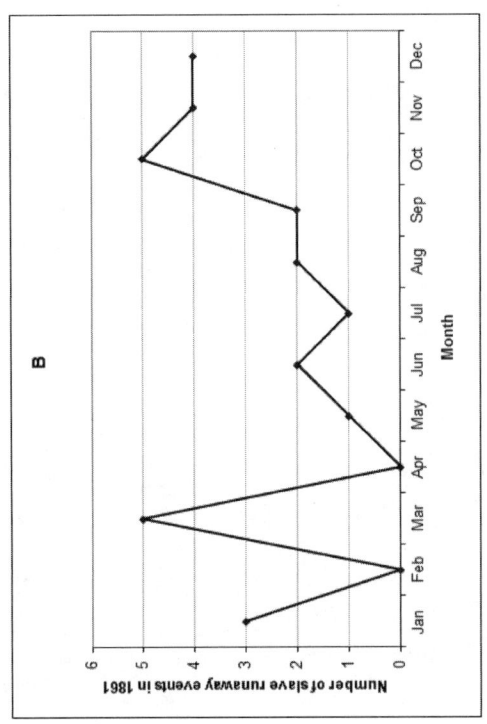

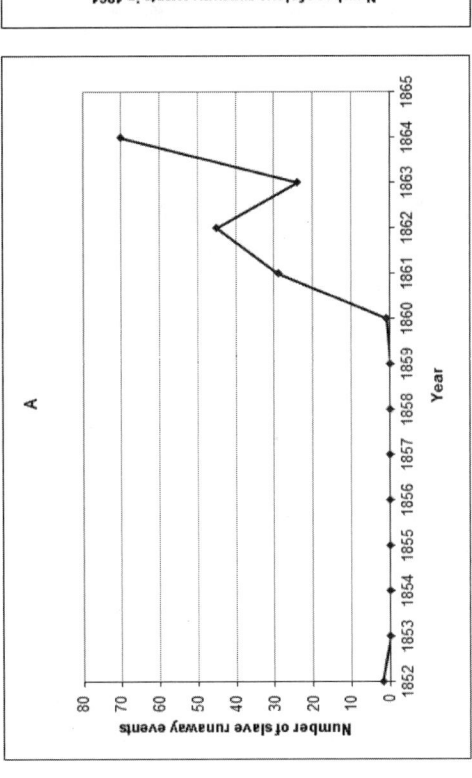

FIGURE 8.6. The slave-run events extracted from *Daily Dispatch* articles: (A) numbers of events per year, (B) numbers of events per month in 1861.

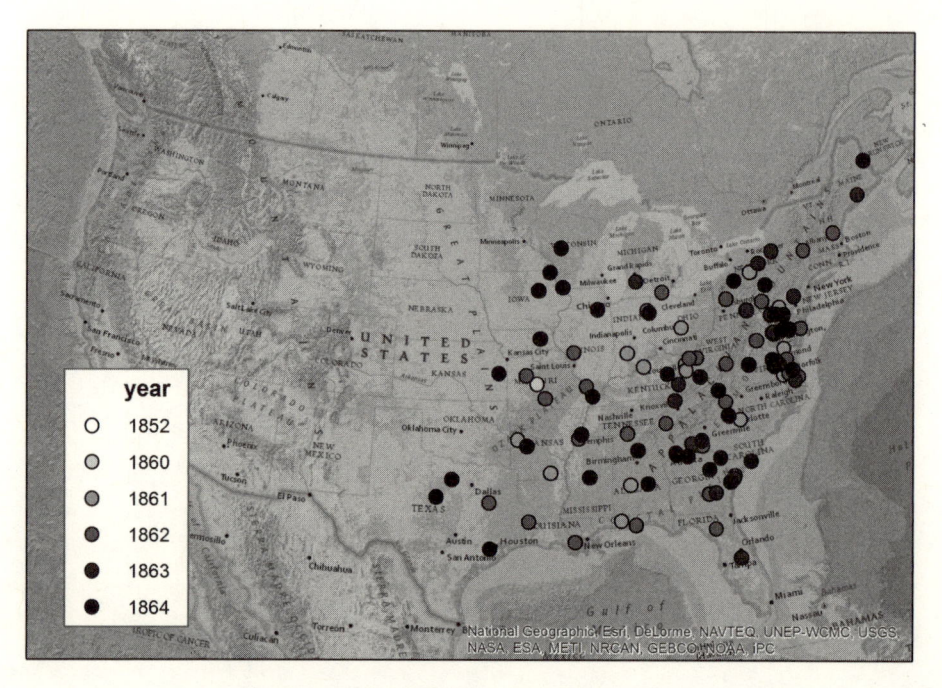

FIGURE 8.7. A preliminary map of slave-run events.

spatially referenced in the shape-file. Five locations referenced to cities in Oregon, California, Utah, and Arizona were determined to be problematic and therefore removed from the map. While the spatial referencing result is still preliminary and demands a careful evaluation, the general spatial pattern suggests the announcements of runaway slaves appear to be widespread in the North, South, and Midwest. Temporal patterns seem obscure.

Another event search seeks reports of infantry movements. *Dyer's Compendium* documents infantry movement events of the Union armies. The *Daily Dispatch* centers mostly on Confederate forces but also reports events involving Union armies. Figure 8.8 shows the preliminary map for all extracted events of infantry movement. A careful evaluation of the spatial referencing results is deemed necessary for these point locations, especially locations in the West. While some of these locations may be problematic, the general trend is likely to hold: Union infantry movements are documented densely in the North and Midwest in *Dyer's Compendium*.

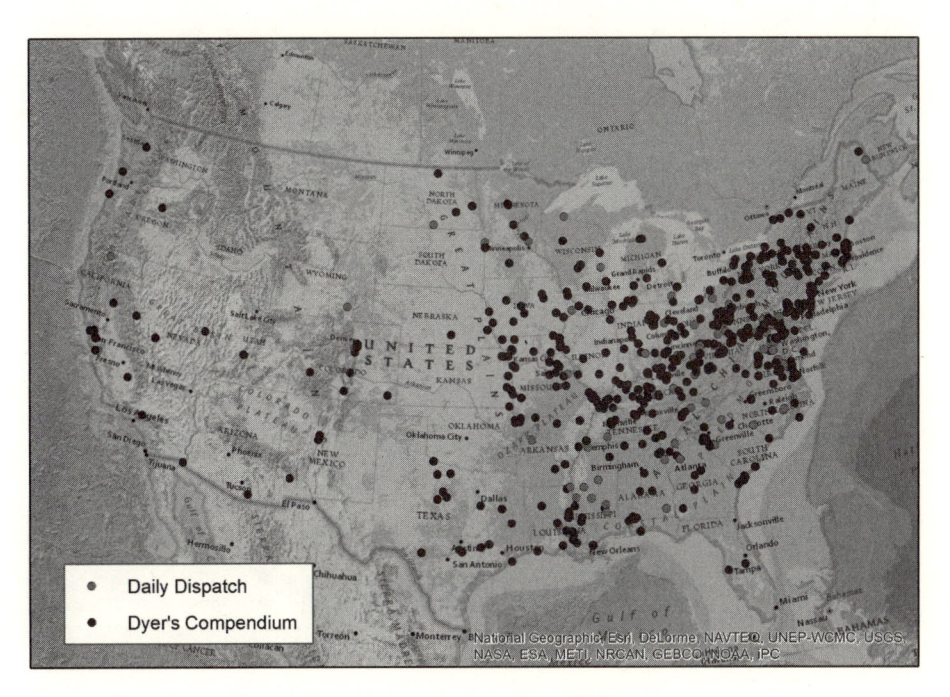

FIGURE 8.8. Reports of infantry movements from both the
Daily Dispatch and *Dyer's Compendium.*

The *Daily Dispatch* adds reports of Confederate infantry movements in the South. Infantry movements in Mississippi, for example, are mostly reported in the *Daily Dispatch.* Events of Union infantry movements in *Dyer's Compendium* appear in a northeast to southwest elongated cluster from New York and Philadelphia through places in West Virginia, Cincinnati, Kentucky, Arkansas, and Louisiana to Baton Rouge and New Orleans. Linear clusters also appear between Iowa and Arkansas as well as Wisconsin to Missouri.

An overlay of slave-run events and infantry movement events suggests that most if not all slave-run events reported in the *Daily Dispatch* are in proximity to Union infantry movements documented in *Dyer's Compendium* (figure 8.9). While no definite conclusion can be drawn without a comprehensive assessment of spatial referencing accuracy for all locations as well as the contents of individual events, a preliminary inspection indicates a high likelihood of spatial correlation between Union infantry

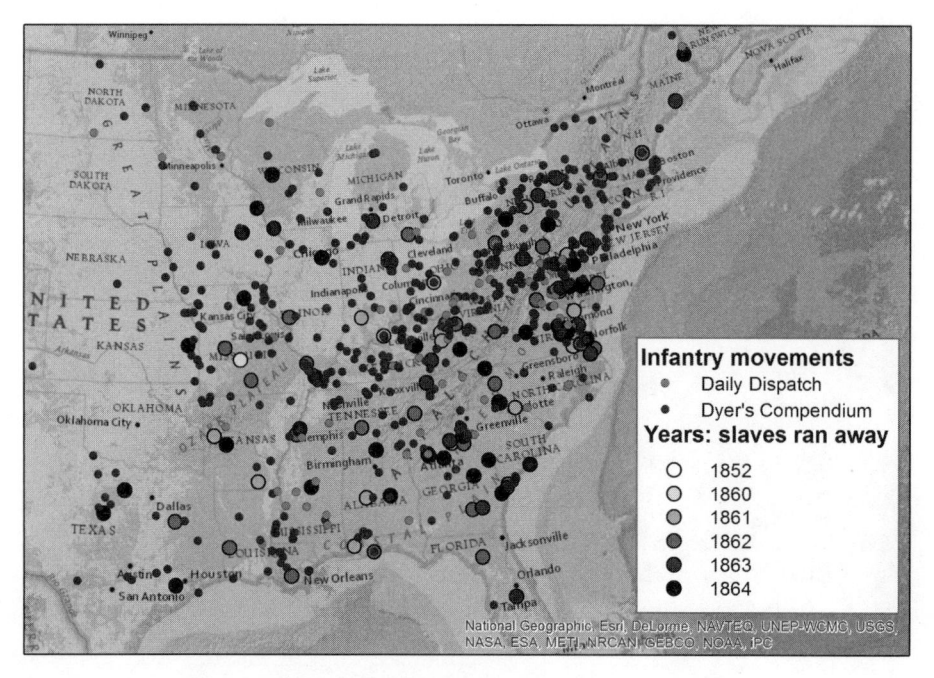

FIGURE 8.9. Events of slave run and infantry movements recorded in *Dyer's Compendium* and the *Daily Dispatch*.

movements and a slave running away. Temporal correlation is also being assessed for possible interactions between the two event types. The use case demonstrates the potential of extracting events from two independent sources, selecting events of interest, and relating events in space and time for new insights to connect events and build spatial narratives.

CONCLUDING REMARKS

This research aims to demonstrate the potential of GIS as a platform for spatial narrative generation. GIS technology traditionally has been an enabling tool for mapping, spatial analysis, and spatial modeling. Much work on temporal GIS has examined approaches to modeling and reasoning about spatial and temporal information. Spatial and temporal aggregation is common in GIS with data aggregated to enumeration units in space and time, such as population per census tract or monthly crime incidents. While temporal GIS research has led to various event-based data models,[26] narrative-based approaches for spatiotemporal data modeling and analysis are uncommon. Narrative-based approaches aggregate spatial information over time to build narratives of journeys or discourses in space and time. Geospatial lifelines present an example of building narratives along someone's experiences in the environment over time,[27] which has gained great popularity with GPS tracking technology and GIS advances in displaying paths in a three-dimensional space-time cube.[28] There are also geonarratives that record textual descriptions of personal journals with locations.[29] Departing from the existing published works, this research defines atomic events and extracts events in four basic elements of action verb, actor, location, and time from textual documents, selects events of interest to assemble GIS data of events, and constructs spatial narratives by spatially and temporally connecting events for historical insights.

This chapter outlines the three main procedures in identifying events and compiling narratives. This research assumes that an action verb marks the occurrence of an event (mostly atomic events). We first developed strategies to identify action verbs as well as nouns that are semantically equivalent. For each sentence with an action verb, procedures are then applied to identify nouns of agents and locations. Proper nouns are evaluated for place names and then are matched with entries in gazetteers to determine the geographic coordinates (i.e., longitudes and latitudes) of these places. Three rules of thumb provide the guidelines for place name disambiguation to determine the most likely location of a place. These rules give preference to the contextual information about geographies (e.g., states, counties, cities, and towns), places closer to the previous location than places further away, and places with larger population sizes. Temporal referencing utilizes chronological time to note explicit time markers that relate events to calendar time or clock time. When absolute time is unavailable or implicit, references to temporal relations can facilitate the ordering of deistic time expressions for mapping and analysis.

An example is provided here to demonstrate the possibilities of such a narrative GIS that can streamline the process of mapping events and narratives in common spatial and temporal frameworks to examine the propagation of events in space and time and from the propagation to investigate insights in which correlations and connections among events of different types or different agents may lead to previously unknown chain effects or events of triggers/drivers that may have changed the course of conviction and lead to initiatives or novel interpretations of historical events by interactively cross-referencing texts and maps. While the example illustrates promising potential for narrative generation, the example shows that the test bed is still in its infancy. With the massive amount of input data, inspection of every event quadruple is impractical. Automatic algorithms are needed to identify possible errors in event extraction, actor determination, spatial referencing, and temporal referencing. Atomic events serve as the blocks which historians can manipulate to elucidate embedded meanings for narrative generation. The test bed and methods in this research present a step forward to fully realize transformation of texts to spatial narratives in GIS and provide a platform which connects rich narratives in historical documents and spatial ramifications of historical narratives in geography.

ACKNOWLEDGMENTS

This material is based upon work supported by the National Science Foundation under grant OCI 0941501. Any opinions, findings, and conclusions or recommendations expressed in this material are those of the author and do not necessarily reflect the views of the National Science Foundation.

NOTES

1. Ian N. Gregory and Richard G. Healey, "Historical GIS: Structuring, Mapping and Analysing Geographies of the Past," *Progress in Human Geography* 31, no. 5 (2007), 638–653. Anne Kelly Knowles and Amy Hillier, *Placing History: How Maps, Spatial Data, and GIS Are Changing Historical Scholarship* (Redlands, Calif.: Esri Press, 2008).

2. John Lewis Gaddis, *The Landscape of History: How Historians Map the Past* (New York: Oxford University Press, 2002).

3. Roberto Franzosi, *Quantitative Narrative Analysis,* Quantitative Applications in the Social Sciences, ed. Tim F. Liao (Los Angeles: SAGE Publications, Inc., 2010).

4. Janet M. Box-Steffensmeier and Bradford S. Jones, *Event history Modeling: A Guide for Social Scientists* (Cambridge: Cambridge University Press, 2004).

5. Dylan Balch-Lindsay and Andrew J. Enterline, "Killing Time: The World Politics of Civil War Duration, 1820–1992," *International Studies Quarterly* 44, no. 4 (2000), 615–642.

6. *Dyer's Compendium* is available at http://www.civilwararchive.com/regim.htm (accessed December 9, 2013).

7. The *Richmond Daily Dispatch* is available at http://www.perseus.tufts.edu/hopper/collection?collection=Perseus:collection:RichTimes (accessed December 9, 2013).

8. The U.S. National Historical GIS project is available at https://www.nhgis.org/ (accessed December 9, 2013).

9. Ian Gregory, Chris Bennett, Vicki Gilham, and Humphrey Southall, "The Great Britain Historical GIS Project: From Maps to Changing Human Geography," *Cartographic Journal* 39, no. 1 (2002), 37–49. Humphrey Southall, "Rebuilding the Great Britain Historical GIS, Part 1: Building an Indefinitely Scalable Statistical Database," *Historical Methods* 44, no. 3 (2011), 149–159.

10. O. W. A. Boonstra, P. K. Doorn, and L. Schreven, "Towards a Historical Geographic Information sSystem for the Netherlands (HGIN). Reports on National Historical GIS Projects," *Historical Geography* 33 (2005), 134–158.

11. Peter K. Bol, "GIS, Prosopography and History," *Annals of GIS* 18, no. 1 (2012), 3–5.

12. The Electronic Cultural Atlas Initiative is available at http://www.ecai.org/ (accessed December 9, 2013).

13. The Arts and Humanities Data Service is available at http://www.ahds.ac.uk/ (accessed December 9, 2013).

14. The Council of European Social Science Data Archives are available at http://www.cessda.org/index.html (accessed December 9, 2013).

15. Natural Language Toolkit is available at http://www.nltk.org/ (accessed December 9, 2013).

16. Humphrey Southall, Ruth Mostern, and Merrick Lex Berman, "On Historical Gazetteers," *International Journal of Humanities & Arts Computing* 5, no. 2 (2011), 127–145. Ruth Mostern, "Historical Gazetteers: An Experiential Perspective, with Examples from Chinese History," *Historical Methods* 41, no. 1 (2008), 39–46.

17. All input documents in this study are in plain English as opposed to markup languages like XML. These input documents contain no tags.

18. The current list of "stop words" include months and days of the week, people titles, and organization titles.

19. Claire Grover, Richard Tobin, Kate Byrne, Matthew Woolland, James Reid, Stuart Dunn, and Julian Ball., "Use of the Edinburgh Geoparser for Georeferencing Digitized Historical Collections," *Philosophical Transactions of the Royal Society A: Mathematical, Physical & Engineering Sciences* 368, no. 1925 (2010), 3875–3889. Judith Gelernter and Nikolai Mushegian, "Geo-parsing Messages from Microtext," *Transactions in GIS* 15, no. 6 (2011), 753–773.

20. David Nadeau and Satoshi Sekine, "A Survey of Named Entity Recognition and Classification," *Journal of Linguisticae Investigationes* 30, no. 1 (2007), 3–26.

21. *Precision* is the fraction of the retrieved named entities out of all retrieved entities correctly identified; that is, the fraction of place names identified is indeed place names. *Recall* is the fraction of the retrieved relevant entities out of all relevant entities in the text of interest is correctly identified; that is, the fraction of place names correctly identified out of all place names in the text.

22. Jochen L. Leidner and Michael D. Lieberman, "Detecting Geographical References in the Form of Place Names and Associated Spatial Natural Language," *SIGSPATIAL Special* 3, no. 2 (2011), pages. Daniel W. Goldberg, "Advances in Geocoding Research and Practice," *Transactions in GIS* 15, no. 6 (2011), 5–11.

23. James F. Allen, "Maintaining Knowledge about Temporal Intervals," *Communications of the Association for Computing Machinery* 26, no. 11 (1983), 832–843.

24. WordNet is available at http://wordnet.princeton.edu/ (accessed December 9, 2013).

25. *Negro* is the common word used in the time period, so this word is regretfully but necessarily used for the search for events related to black men escaping from slavery. In the text that follows, we use the word *slave* instead in discussion.

26. D. J. Peuquet and N. Duan, "An Event-based Spatiotemporal Data Model (ESTDM) for Temporal Analysis of Geographical Data," *International Journal of Geographical Information Systems* 9, no. 1 (1995), 7–24. J. McIntosh and M. Yuan, "Assessing Similarity of Geographic Processes and Events," *Transactions in GIS* 9, no. 2 (2005), 223–245. Michael Worboys and Kathleen Hornsby, "From Objects to Events: GEM, the Geospatial Event Model," in *Proceedings of the International Conference of Geographic Information Science 2004, Adelphi, Md., October 20–23, 2004*, ed. Max Egenhofer, Christian Freksa, and Harvey Miller, 327–343 (New York: Springer, 2004).

27. David M. Mark and Max J. Egenhofer, "Geospatial Lifelines," in Oliver Gunther, Timos Sellis, and Babis Theodoulidis, eds., *Integrating Spatial and Temporal Databases* (Dagstuhl Seminar Report No. 228, Schloos Dagstuhl, Germany, 1998).

28. Mei-Po Kwan, "GIS Methods in Time-Geographic Research: Geocomputation and Geovisualization of Human Activity Patterns," *Geografiska Annaler Series B: Human Geography* 86, no. 4 (2004), 267–280.

29. Mei-Po Kwan and Guoxiang Ding, "Geo-Narrative: Extending Geographic Information Systems for Narrative Analysis in Qualitative and Mixed-Method Research," *Professional Geographer* 60, no. 4 (2008), 443–465.

9

WARP AND WEFT ON THE LOOM OF LAT/LONG

WORTHY MARTIN

Human agency and action occurs in space-time and is conceptualized in numerous social and cultural aspects. To focus on one such aspect can provide a discipline through which a coherent narrative can emerge. Fundamental to this chapter and, indeed, this whole volume, is that electronic, internet-accessible resources enable the creation and dissemination of narratives that span several of those social and cultural aspects, with each being a strand in a larger story with points of contact among the various strands. That assumed fundamental capability questions which disciplines can facilitate the creation of new forms of coherence for scholars and the general public approaching the varied resources. An additional promise of presentation via electronic media is that the resources will have affordances for individualized coherence.

The loom of latitude and longitude provides a framework in which to create deep maps that yield access paths, through predesigned strands and through individualized selections, to spatial narratives. The temporal dimension then is available for animating those spatial narratives, thus bringing given aspects of human agency and activity to life.

In the remaining sections of this chapter, four case studies will be presented to show a range of deep maps through which a variety of spatial narratives are expressed. In each case, crucial design decisions and alternatives are discussed.

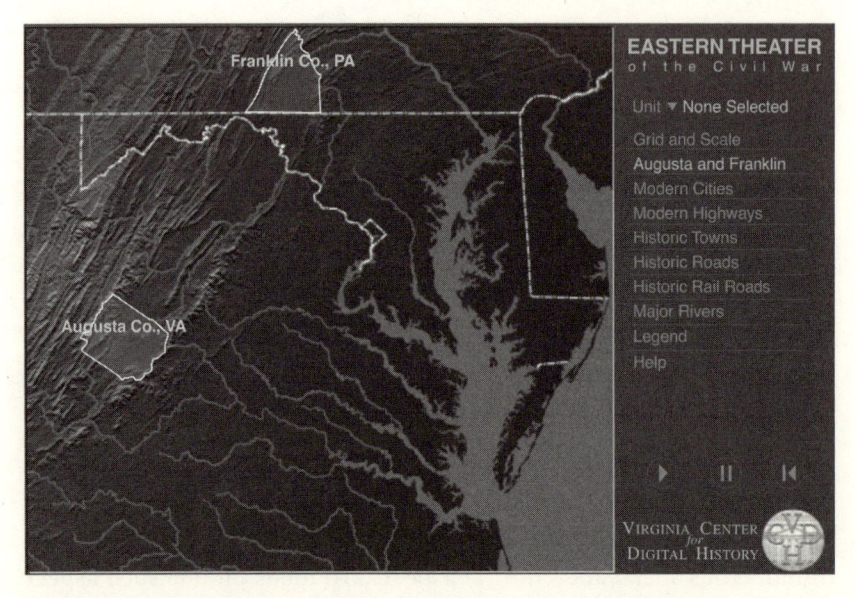

FIGURE 9.1. Two counties in the same geological and ecological setting, but on opposite sides in the U.S. Civil War.

THE VALLEY OF THE SHADOW

The counties of Franklin (in Pennsylvania) and Augusta (in Virginia) share a common geological and ecological setting (see figure 9.1). In the agrarian nineteenth century, this commonality implied cultural similarity. However, the political boundaries drawn between the counties lead to separation in the U.S. Civil War. The narratives developing in this spatial commonality with the coming of the political separation are the focus of the Valley of the Shadow project.[1] The documentation for those narratives comes from contemporary newspaper articles, personal diaries, and correspondence (as transcribed by the project).

Because of the military conflict, many of the personal narratives involve participation in army units. The two counties each had three army units, namely, for Franklin—the 107th Pennsylvania, the 126th Pennsylvania, and the 16th Cavalry—and for Augusta—the 5th Virginia Infantry, the 1st Virginia Cavalry, and the Staunton Artillery. For this particular aspect, a spatial narrative for those units is provided by considering the

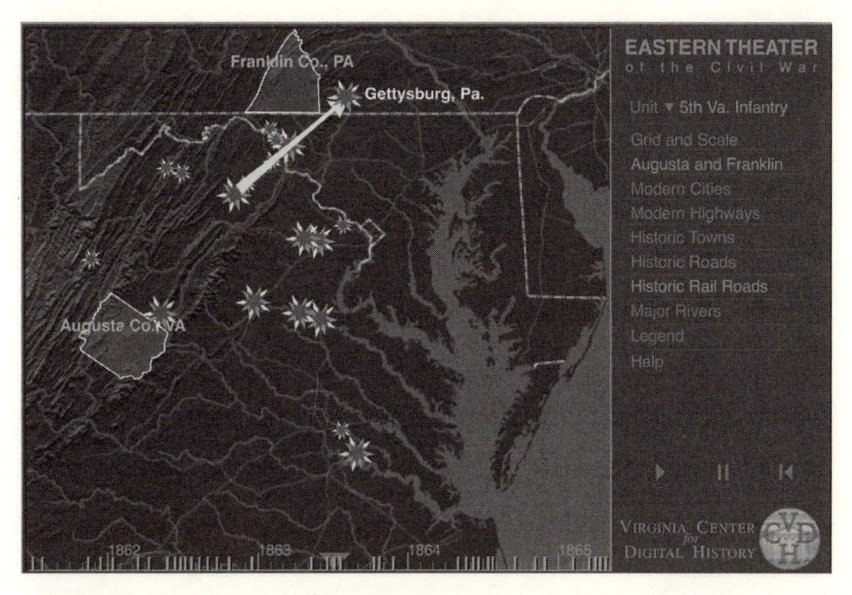

FIGURE 9.2. Progression of battles and engagements for the Fifth Infantry unit from Augusta County, Virginia.

battles and engagements that the units joined. These six strands of the narrative play out in the Eastern Theater, thus the geographical context of the display in figure 9.2 is that of Virginia, Maryland, and Pennsylvania. The progression of battles and engagements calls for a timeline and the capability of animation. As an example, the Fifth Virginia Infantry was involved in eighteen battles and engagements before it joined the battle at Gettysburg on July 1, 1863. Figure 9.2 shows the animation of the progression for the Fifth Virginia Infantry paused on July 1, 1863. The distribution of the prior battles and engagements is indicated by an icon for each. The transition from the last of those, Winchester, Virginia, to Gettysburg is shown as an arrow connecting the two locations.

Since a military unit might have had a series of battles and engagements in a fairly restricted area (relative to the scale of the theater map), occasionally a location is listed as a "campaign." Selecting that campaign icon then presents a higher resolution inset with a separate timeline. On June 15, 1864, the 107th Pennsylvania began a series of battles and engagements

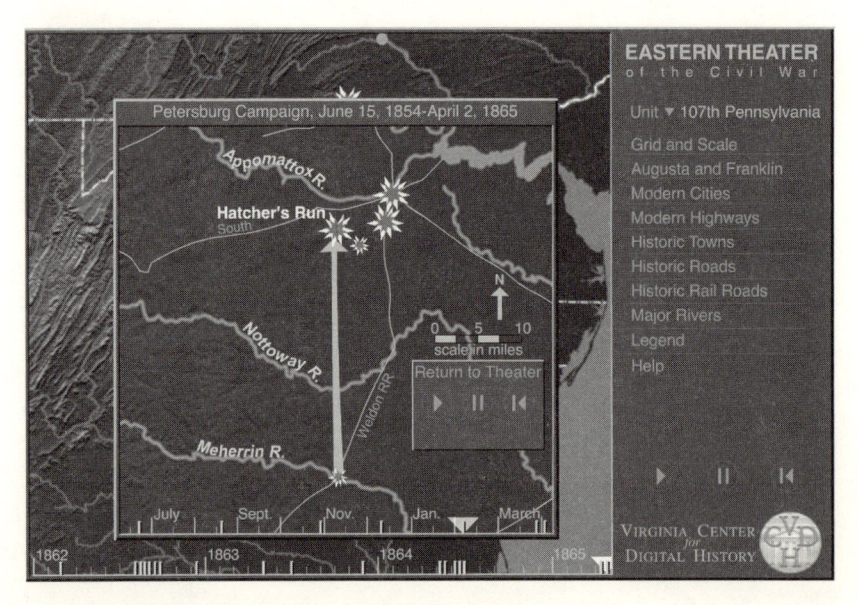

FIGURE 9.3. Detail of a campaign within a restricted area for the Pennsylvania 107th.

in the vicinity of Petersburg, Virginia, and the unit's movement to join the Hatcher's Run battle is depicted in figure 9.3.

Each battle and engagement is linked to a textual record that provides detailed facts, such as battle commander, a summary of the events at the battle, precise location, weather details and description, and links to transcriptions of official records.[2] At the bottom of each map display a timeline indicates the specific date (triangle). The temporal distribution of the battles and engagements for the Fifth Virginia Infantry is presented via highlighted tick marks for each corresponding date.

Additional geographical context for the battles and engagements narratives are provided by layers that can be displayed. The network of light gray lines evident in figure 9.2 (but not in figure 9.1) are the railroads of the 1860s resulting from the selection of the "historic railroads" layer. As one might well expect, almost all of the battles[3] are at locations served by railroads of the time. Included in the optional layers are other geographical contexts, such as "modern highways."

Through the deep map of figure 9.2, aspects of the intertwined narratives of the fathers, sons, and brothers of the families in Augusta and Franklin counties are presented in the eastern theater of the American Civil War and through that geographical context related to modern lives. However, those narratives are presented in terms of groups, that is, the military units. In the next section, an episode that has reverberated through the entire history of America is addressed to focus on the individuals involved.[4]

SALEM WITCH TRIALS

The cultural relationships of the participants in what today we call the Salem witch trials are richly documented in legal documents, personal correspondence, and secondary scholarship on those primary materials. The mission of the Salem Witch Trials Documentary Archive and Transcription Project is to make those primary materials available to scholars and the general public.[5] The spatial narratives of the sequence of accusations and trials have generated substantial textual scholarship. Here a deep map display of a form of that spatial narrative will be discussed.

The initial consideration for this deep map is how the social interactions evident in the court documents are to be placed in a geographical context. Certainly, each case occurred in a "court" and that location is of social significance. However, that single location does not relate the richness of the interpersonal events transpiring in each case through accusations. To present aspects of those interactions, the people and their roles are crucial. Over the primary period of these trials, February through November 1692, each person was part of a specific household. Since one's family was an important factor in how one was involved in these episodes, the household serves as a social nexus and provides a relevant geolocation for each member of the household, namely, the primary family residence. As part of the amazing amount of documentation of these trials, there is a nineteenth-century map[6] indicating the location of almost every household in 1692.[7] In this way, each person can be located in a deep map along with their household members.

The deep map that conveys the spatial narrative of court cases through the individuals who make accusations and those accused is shown in fig-

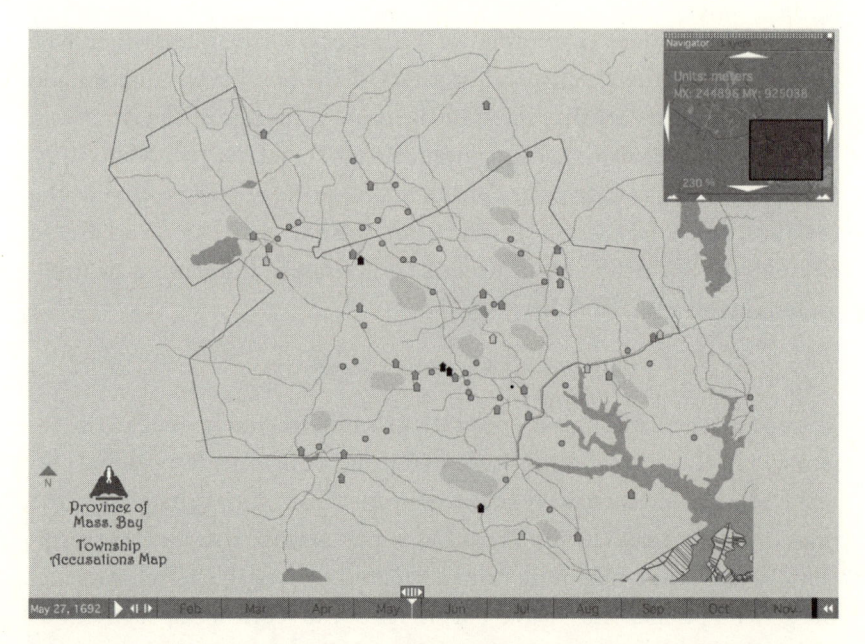

FIGURE 9.4. Map of Salem Village with households indicated. Highlighted are the households of court case participants (accusers and accused) for May 27, 1692. The relationship to the whole Province of Massachusetts Bay is provided by the insert and a timeline of the crucial months of the "trials" is given.

ure 9.4. The geographical framework in which the strands of that narrative are woven is the Province of Massachusetts Bay. The households known to have existed in 1692 are given at the location of the primary building of the household using two icons. The house icon is used for households in which at least one individual was a participant (accuser or accused) in a court case. The circle icon is used for households in which no individual was a participant. Immediately presented by the deep map is a narrative strand that suggests the spatial distribution and intermixing of court case participants.

The narrative is expanded to include the temporal progression of the court cases through a timeline, with a "play" option to convey the dynamics of the accusations by highlighting the households of the participants in court cases each day of the overall period. In figure 9.4, the animation is paused on May 27, 1692, and the view restricted to the southeast part

Warp and Weft on the Loom of Lat/Long 209

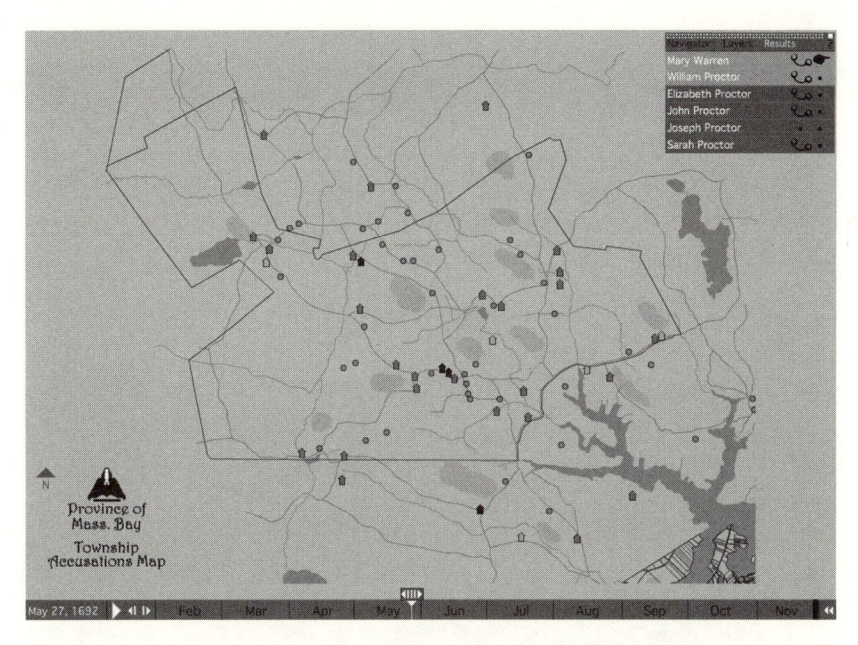

FIGURE 9.5. The individuals in the household of John Proctor (lowest highlighted location) are shown in the inset. For each there is an indication of whether they accused someone or were accused. Highlighting in the inset indicates that the action took place on May 27, 1692.

of the region (as indicated by the darker gray square inside the pan/zoom control box in the upper right of the display). In the main display, there are four households highlighted indicating participants in court cases that day.

To present the personal strand of the narrative, it is possible to select a specific household and a list of the household members (direct family and others associated with the family, such as servants) replaces the pan/zoom control. The result of selecting the John Proctor household (the lowermost highlighted household in figure 9.4) is shown in figure 9.5. The Proctor household had six members: John Proctor, Elizabeth Proctor, Sarah Proctor, Joseph Proctor, William Proctor, and Mary Warren. All but Joseph Proctor were accused during the trials period (indicated by the handcuff icon) and Mary Warren accused someone else (indicated by the pointing hand icon). The highlighting on Mary Warren's accusation

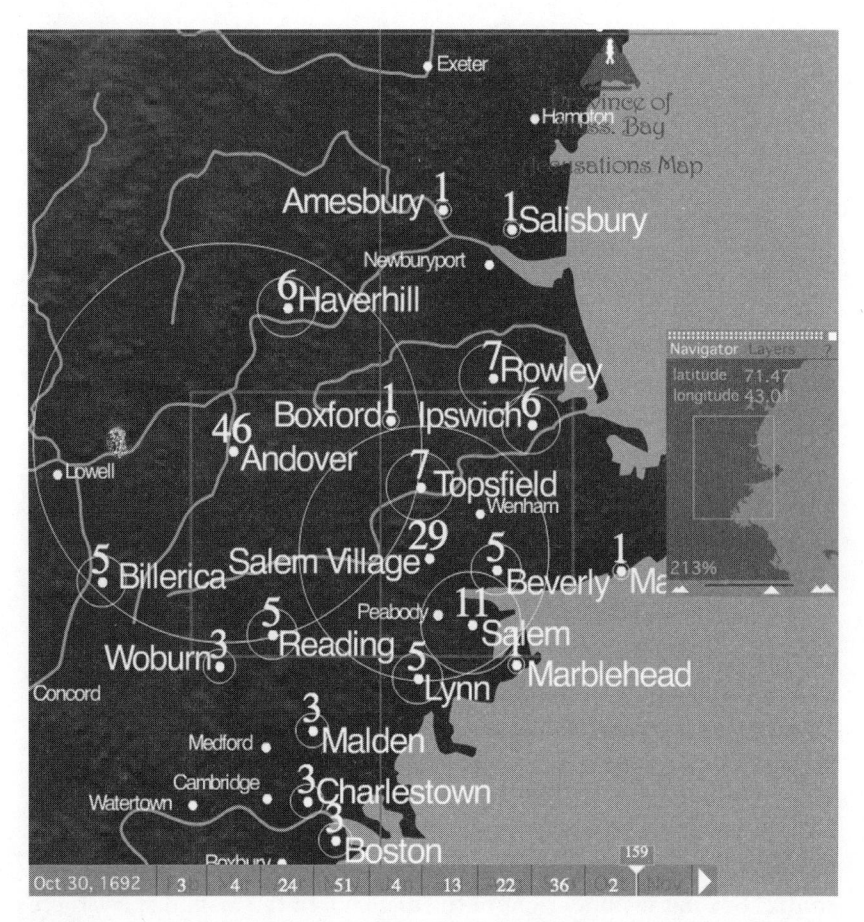

FIGURE 9.6. The region of the Massachusetts Bay Province with accumulative totals by township of accused individuals as of October 30, 1692.

and the accusation of William Proctor indicate that those court actions took place on the selected day. The depth of the map is extended by the ability to access a brief (and possibly extended biography) of each person by selecting his or her name in the list. Indeed, the transcriptions of the entire court records[8] are then available through the deep map.

The narrative of aggregate data across the Massachusetts Bay region is available via a separate map. In figure 9.6, the geographic context for the region is shown with the major towns and villages labeled. As in the map presented in figures 9.3 and 9.4, there is a pan/zoom control box and a timeline that enables animation. In this case, the temporal progression is of the cumulative total of accusations made at each location. The specific totals are shown at each location and a graphical indication of the magnitude of each total is depicted as a circle centered at the town/village with a radius proportional to the total. The timeline shows the monthly total aggregated across the region in the block for each month.

The deep map in figure 9.6 immediately conveys an overall aspect of the court case spatial narratives that is surprising to most, namely, that the phenomena over this specific time period was not contained in the area immediately around Salem Village. Indeed, Andover had 50 percent more accusations than Salem Village did (relative to Salem Village's total). While this deep map is of aggregate information, it is based on detailed documentation of individuals, their activities, and agencies. In the next section, the spatial narratives are of a people for whom the cultural relationships are at best only vaguely implied in an extensive archaeological record (as it is presently known and understood).

CHACO CANYON

Chaco Canyon has one of the longest histories of archaeological research in the Americas, with initial professional studies dating to 1887. Between 800 and 1100 CE, a period of unprecedented construction resulted in massive masonry buildings, or great houses, along a five-mile stretch of Chaco Wash. For almost 125 years, archaeologists have struggled to understand the social transformation that took place in Chaco culminating in this astounding florescence of human activity and agency. The Chaco Research Archive (CRA)[9] has been created to enable new possibilities for synthetic research and for interpretation over temporal and spatial scales not practical before with regard to the Chaco Canyon area of northwest New Mexico.

A separate strand of human activity and agency is formed by the archaeological expeditions to Chaco Canyon that excavated and aggregated primary evidence of the earlier occupation. That archaeology strand then transforms into an archivist strand with the curation of the primary evidence into archive, repository, and museum collections.

The threat of nonarchaeological expeditions into Chaco Canyon led in 1907 to a national monument designation[10]—now also a U.N. Educa-

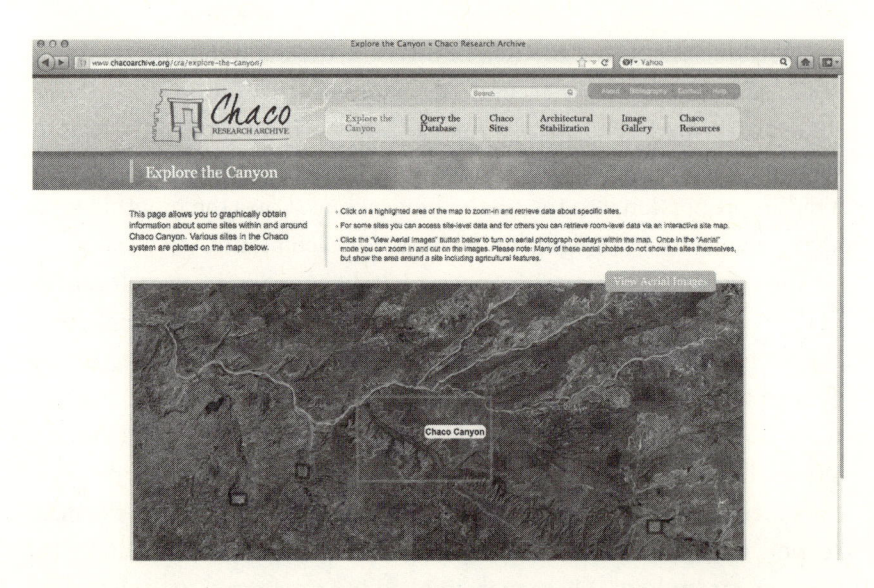

FIGURE 9.7. Chaco Research Archive web context and geographical setting.

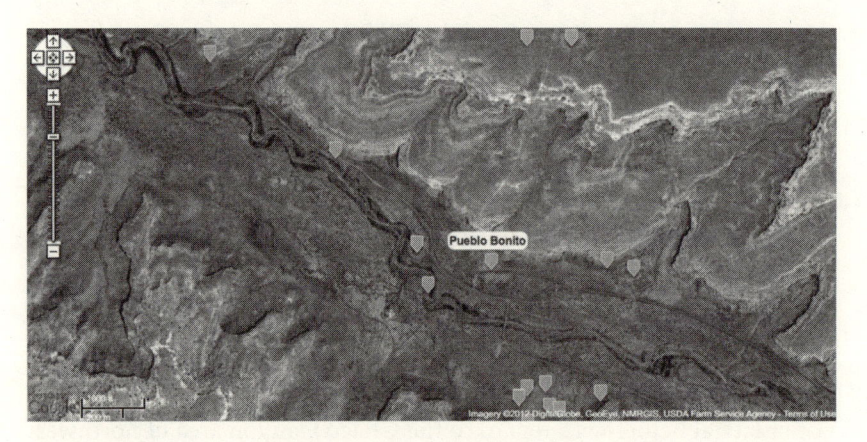

FIGURE 9.8. Primary archaeological sites indicated in geographical context, one with the label available via a mouse-over.

tional, Scientific, and Cultural Organization world heritage site. Thus, for over a hundred years, the National Park Service has had custodial and curatorial control of the land and cultural heritage, creating another strand of human activity and agency impacting and being impacted by the canyon.

Warp and Weft on the Loom of Lat/Long 213

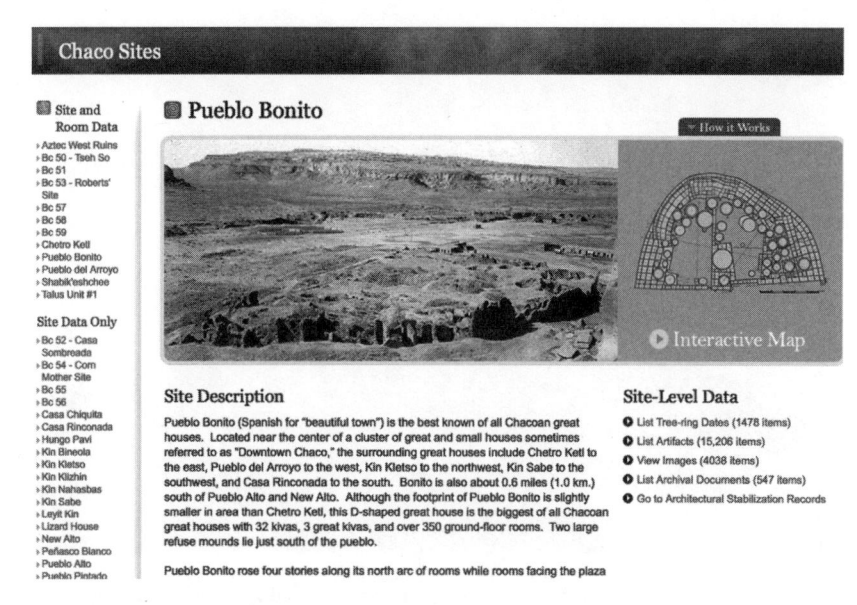

FIGURE 9.9. Nexus of information about the specific site, Pueblo Bonito.

The CRA project approaches its mission to enable synthetic research and interpretation by presenting these stands interwoven in deep maps and spatial narratives that make the archived evidence[11] accessible to contemporary scholars and the general public. In the remainder of this section, I will present and discuss the specific forms of deep maps and spatial narratives that create the framework in which the archived evidence is brought to light.

An initial level of the deep map for Chaco Canyon is shown in figure 9.7. The rectangle (here with a mouse-over label and light gray highlighting shown) circumscribes the official national park area. After two steps of zoom (the second also prompted by a labeled rectangle) the display appears as in figure 9.8 presenting the area often called downtown Chaco. Here the primary archaeological sites for which CRA has archival information are shown with actionable icons. In addition, the most-studied great house, Pueblo Bonito, is labeled (in the captured display via mouse-over of its icon). The result of selecting Pueblo Bonito is displayed in figure 9.9. The major strands are evident with the options for primary data—tree ring data and artifact list—and curatorial and custodial responsibility—

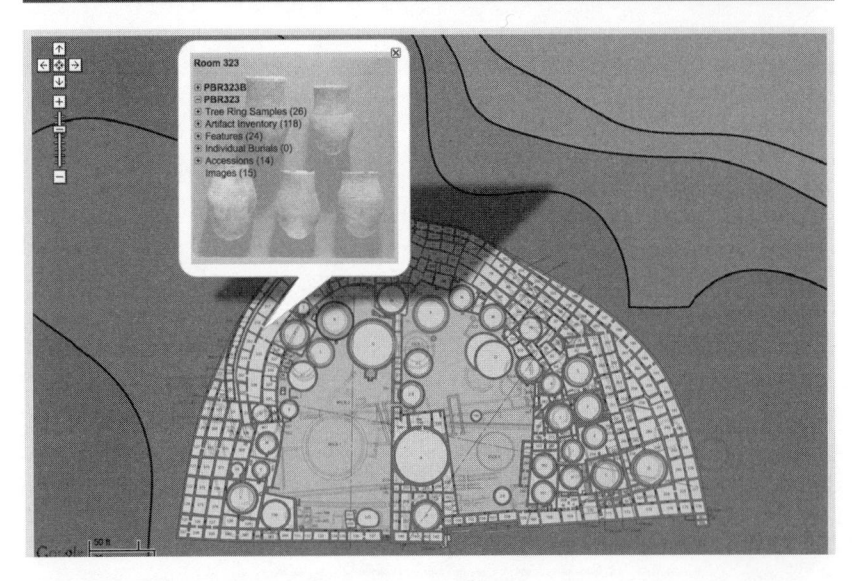

FIGURE 9.10. Archival information for room 323 revealed for deeper access via a plan view of the great house, Pueblo Bonito, incorporating inscriptions from original archaeological drawings.

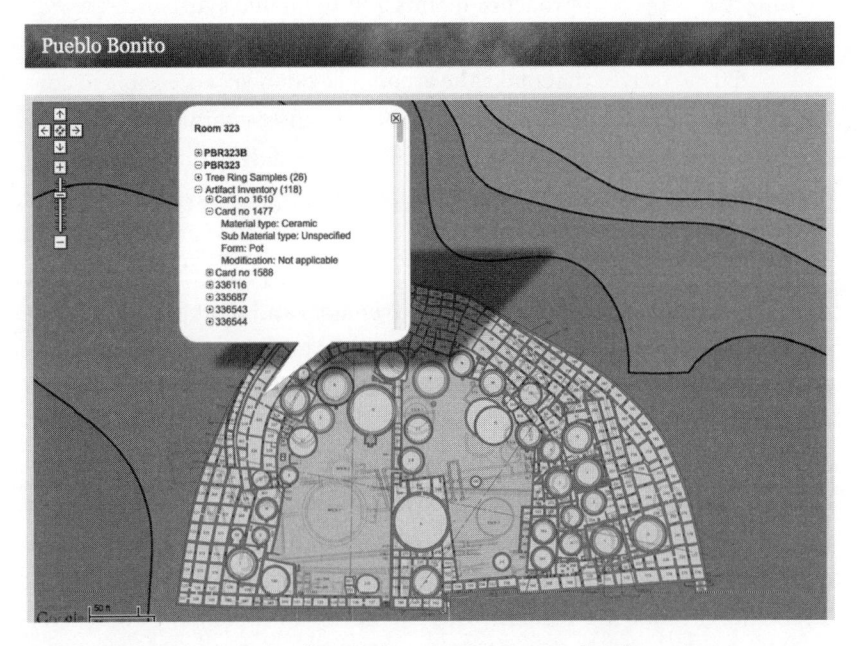

FIGURE 9.11. Preview of archival information associated with room 323.

Artifact List for Room 323 (PBR323)					
Specimen Number	Field Catalog Description	Material Type	Sub Material Type	Form	Modification
Card no 1610		Unspecified			Unspecified
Card no 1477		Ceramic	Unspecified	Pot	Not applicable
Card no 1588		Ceramic	Unspecified	Pot	Not applicable
336116		Ceramic	Clay	Unspecified	Unspecified
335687		Shell	Olivella	Bead	Worked
336543		Ceramic	Unspecified	Pot	Not applicable
336544		Ceramic	Unspecified	Pot	Not applicable
335700		Shell	Haliotis	Pendant	Worked
335256		Wood	Unspecified	Unspecified	Worked
335270		Wood	Unspecified	Tablet	Worked
335254		Wood	Unspecified	Stick	Worked
335261		Wood	Unspecified	Unspecified	Worked
335248		Wood	Unspecified	Unspecified	Unspecified
335632		Unspecified	Unspecified	Concretion	Worked
335401		Mineral	Unspecified	Paint/Pigment	Worked
335154		Bone	Unspecified	Scraper	Worked
335249		Wood	Unspecified	Unspecified	Unspecified
335047		Bone	Unspecified	Awl	Worked

FIGURE 9.12. Detailed textual presentation of archival information from room 323.

"architectural stabilization records"—while the image collection and archival documents cover all modes of activity. The materials shown are all filtered to be specific to Pueblo Bonito.

This interweaving of the strands becomes evident in a more intellectually refined manner through returning to the geographical context of a plan view of the great house (shown in figure 9.10).[12] Here the deep map is organized around the individual rooms of Pueblo Bonito, with each room providing access to the archival information and primary data associated with the specific room. A major interest for both archaeologists and the general public are the artifacts found in the room. A preview of the artifact inventory can be seen in the geographical context in figure 9.11. Also, available is the complete inventory of artifacts from the room, as shown in figure 9.12.

> Refilling Room 323, Pueblo Bonito. Center front R. 325
> refilled. #28482. Photos by O. C. Havens, 1924.
> NAA judd_ngs_0939
> National Anthropological Archives
> NGS 28482
>
> Commercial use of these images is strictly prohibited without prior written permission from the repository institution.
> Image 14 of 15
>
> CLOSE ✕

FIGURE 9.13. Individual image from catalog of images associated with room 323.

The historical imagery available (at the room level, the site level, the institution level, and as an overall gallery) advances the spatial narratives of several of the strands. Certainly the artifacts and architectural structures are depicted as found and as collected to glimpse the lives of the original occupants. In addition, the archaeologists, laborers and visitors associated with the archaeological expeditions, as well as the excavation processes they used, are documented. Those processes cover the spectrum from physical, for example, digging, cleaning, and reassembly, to the informational, for example, field notebooks and drawings. Finally, aspects of the continuing custodial strand are documented in the images from the

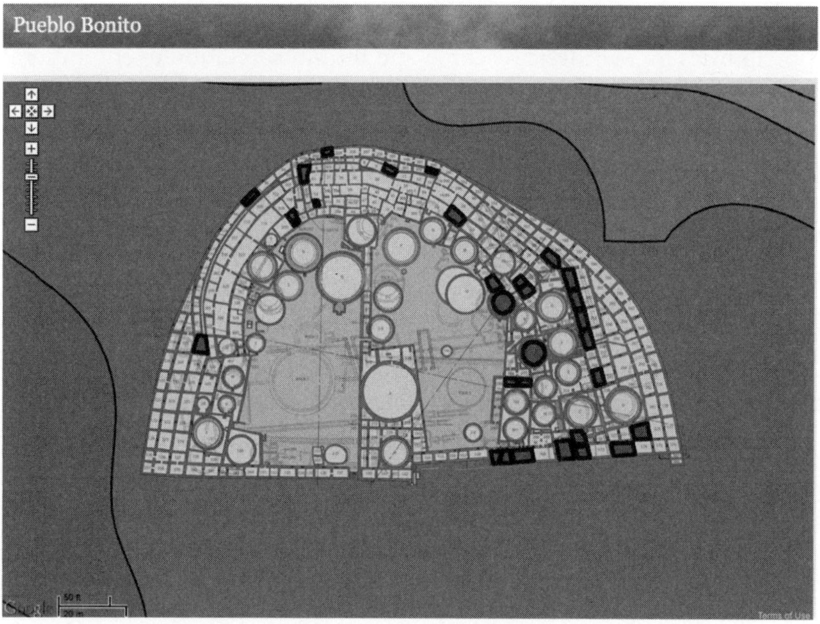

FIGURE 9.14. The plan view of Pueblo Bonito highlighting the rooms in which painted wood artifacts were found.

processes of stabilization and preservation of the architectural structures. The catalog of images associated with room 323 of Pueblo Bonito can be shown as a set of thumbnail images and then individual images can be selected to be seen at a higher resolution as shown in figure 9.13.

This section has presented the deep map as an access mechanism presenting aspects of several spatial narratives. The underlying deep map information structure also allows one to begin with a categorical inquiry then to consider the spatial relationships. For example, by posing a query to the CRA database asking for all artifacts found at Pueblo Bonito that are classified as "CRA material type: wood" in the "Field catalog description" containing "painted," a scholar can get a textual list of the database records for those artifacts. However, the spatial relationships of the rooms in which those artifacts were found can be considered in a plan view display (figure 9.14)[13] in which the associated rooms are highlighted (dark borders). This access to the primary materials from the numerous expeditions is at the heart of the CRA mission to enable new possibilities for synthetic research and for interpretation over temporal and spatial scales not practical before with regard to the Chaco Canyon area of northwest New Mexico. Such possibilities derive from extensive CRA efforts to obtain digital surrogates of original documents and to transcribe, normalize, and aggregate the primary data from those documents. However, further extensive efforts have been required to design and implement the deep maps through which the spatial narratives (often with specific strands highlighted) are presented for scholars and the general public.

In the next section, spatial narratives will be considered for a domain in which myth and history blend, yet in which the activity and agency of individuals and the associated relationships are the focus.

HOMER'S THEATER IN THE *ILIAD*

While she does not use the term, Jenny Strauss Clay argues that Homer establishes a spatial narrative in the main battle books of the *Iliad*,[14] indeed that he chooses to set the Greek encampment in the foreground and orients "stage right" and "stage left" to be from the Greek perspective of looking toward Troy from the beach. In depicting the geographical context for the action of the battle books, extraneous and unjustified detail must be avoided. Thus, the map shown in figure 9.15 is schematic[15]; yet it is set to be consistent with the local orientations expressed in individual battle descriptions and with the regional orientations of the entire epic poem.

The specific books of the *Iliad* that Clay analyzed are given on the right of figure 9.15, with the option to see the poem in either a Greek[16] or English[17] version. Selecting the Greek option for book 15 results in the display of figure 9.16. Because the action in book 15 is focused around the wall constructed by the Greeks and then their encampment among their ships on the beach, the display is zoomed-in and centered in that area. The text of the poem is presented in the lower left, identified by book and line number.

The implicit nature of the expression of the temporal progression of the battles precludes timelines such as presented in the Valley of the Shadow project and the Salem Witch Trials Archive.[18] However, the presentation

Homer's Trojan Theater

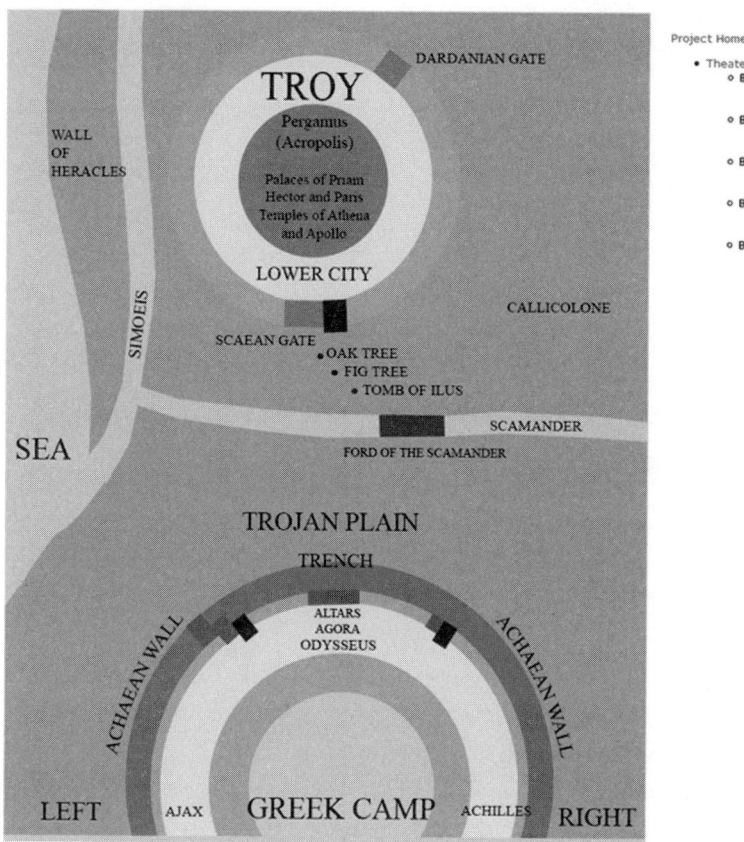

FIGURE 9.15. Schematic interpretation of the theater of action in the *Iliad*.

order of the lines of the poem does provide a type of temporal progression. A timeline-like visual element is included as the vertical line just to the right of the schematic map. The line represents the sequence of textual lines for the complete book selected. The analyzed action events are indicated as tick marks positioned down the vertical line with the distance proportional to the line number. Line-by-line scrolling of the text is available by dragging the triangular icon that begins at the top of the vertical line. The first line of each action event is highlighted (light gray in figure 9.16) in the text box in the lower left,[19] appearing at the top of the box as

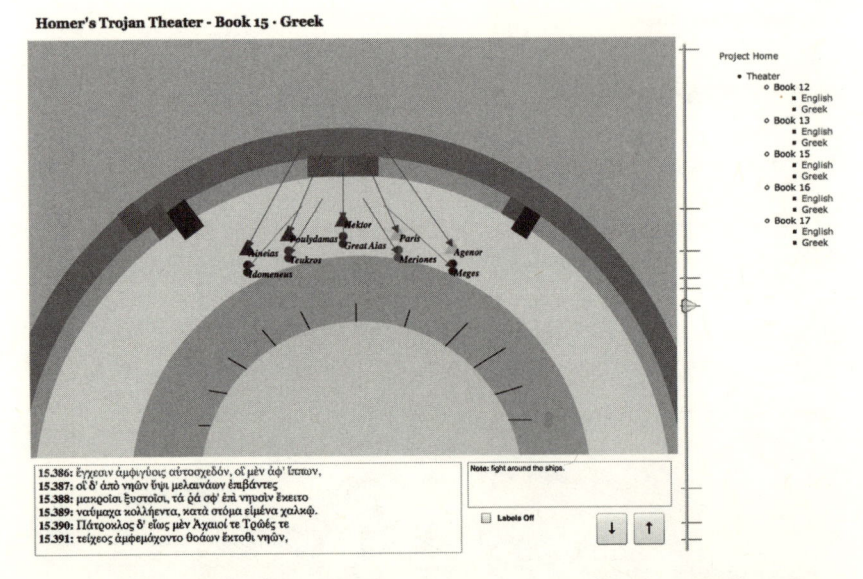

FIGURE 9.16. Depiction of action described in line 386 of book 15, Greek version.

the triangular icon reaches the associated tick mark. It is also possible to scroll directly through the action events using the up and down arrows just to the left of the bottom of the vertical line.

The battle action narrative that unfolds through the presented progression is spatialized with icons for participants and some forms of action. The individuals are associated with two types of icons: triangles (Trojan) and paired disks (Greek). The placement of the icons is, again, very schematic; however, the poem is often explicit about paired combatants and the placement is made consistent with such pairings. Each individual is named.

The underlying information structure has "event" as the primary category. Each event has it beginning line in the specified book, a list of participants, and a spatialized visual element. It was an explicit design decision to maintain a fixed spatial framework (the latitude/longitude in this case) for each book, with options to see the overall context and to see more detailed labeling of landmarks. Holding a spatial context constant for conceptual coherence (in this case, throughout a book of the poem) is crucial for effective spatial narratives. The selection of that spatial context often presents a trade-off concern (as in all cartography) for the quantity and level of granularity of the information displayed. For the *Iliad* battles, this trade-off demands the schematic nature of the spatializing visual and even editorial license in the selection of the list of participants in any one event.

CONCLUSIONS

In the preceding sections, four deep maps have been presented with a discussion of the design considerations at both the "graphic design" level and the underlying information structure level. Latitude and longitude (even as schematized for mythological landscape of the *Iliad*) provide the framework within which spatial narratives can be interwoven to make accessible the full range of human agency and activity. Establishing the framework is crucial to the intellectual accessibility of the full story. However, it is equally crucial that the coherence of each individual spatial narrative is available for scholars and the general public to provide intellectual accessibility to the detailed interconnections within specific social and cultural aspects of the human agency and activity. Finally, the discipline of the spatial framework and of the individual strands of the spatial narratives are combined via electronic media to create affordances for individualized exploration that hold promise for emergent spatial narratives beyond those originally envisioned.

NOTES

1. The Valley of the Shadow project is available at http://valley.lib.virginia.edu/ (accessed August 24, 2013).
2. For example, the Report of Brig. Gen. Lysander Cutler, U.S. Army, commanding Fourth Division is reproduced in *The War of the Rebellion: Official Records of the Union and Confederate Armies*, series 1, vol. 36 (Wilmington, N.C.: Broadfoot Publishing Company, 1997), part 1, reports, serial no. 67, 610.
3. The few engagements that are not on railroad lines are in the mountainous western portions of the theater.
4. It is important to observe that for spatial narratives at the individual person level, the scale of the episode (e.g., hundreds of people over several months versus millions of people over several years) is crucial from both the practical considerations of producing the presentations and documentary evidence and the intellectual considerations of producing coherent presentations that scholars and the general public can assimilate and find engaging.
5. The Salem Witch Trials Documentary Archive and Transcription Project is available at http://salem.lib.virginia.edu/home.html (accessed August 24, 2013).
6. C. W. Upham, *Salem Witchcraft, with an Account of Salem Village and a History of Opinions on Witchcraft and Kindred Spirits* (Boston: Wiggin and Lunt, 1867).

7. Professor Ben Ray, the lead scholar at the University of Virginia for the project, was able to augment the Upham map with several more household locations.

8. P. Boyer and S. Nissenbaum, eds., *The Salem Witchcraft Papers: Verbatim Transcripts of the Legal Documents of the Salem Witchcraft Outbreak of 1692* (New York: Da Capo Press, 1977).

9. The Chaco Research Archive is available at www.chacoarchive.org (accessed August 24, 2013).

10. On March 11, 1907, Chaco Canyon National Monument was created under the authority of the Antiquities Act. On December 19, 1980, the area covered was increased and the designation changed to the Chaco Culture National Historical Park.

11. Here we mean "digital surrogates" of the originals. The surrogates range from scans of documents and photographs (presented in pdf, jpeg, etc.), full text transcriptions of documents (including field notes and personal correspondence), datasets as searchable database tables and as downloadable spreadsheets, and of major importance, normalized value fields (along with the as-written forms).

12. Here the geographical context is changed from satellite imagery to an abstracted map of topographic lines. This serves to eliminate the confusion created by the non-alignment of incidental details (in both the photographic imagery and in the plan view drawing).

13. If the scholar is studying several sites, then the site-level spatial relationships are available through the canyon scale map display, which is not shown here.

14. J. S. Clay, *Homer's Trojan Theater: Space, Vision and Memory in the Iliad* (Cambridge: Cambridge University Press, 2011).

15. The schematic maps used at http://www.homerstrojantheater.org/ (accessed August 24, 2013) and shown in figures 9.15 and 9.16 were adapted from B. Mannsperger, "Das Stadtbild von Troia in der Ilias," in J. Latacz, ed., *Troia: Traum und Wirklichkeit* (Stuttgart: K. Theiss, 2001), 81.

16. D. B. Munro and T. W. Allen, eds., *Oxford Classical Texts: Homeri Opera* (New York: Oxford University Press, 1920).

17. R. Lattimore, *The Iliad of Homer* (Chicago: University of Chicago Press, 1961).

18. No rigorous "total ordering" of the events is available, that is, many events are described without a clear indication of which one began or ended first.

19. Brief "further information" is provided in the box to the right of the primary text box.

CONCLUSION

Engaging Deep Maps

TREVOR M. HARRIS, JOHN CORRIGAN,
AND DAVID J. BODENHAMER

Deep mapping arose out of the imaginary of the spatial humanities, and while it is conceptually rich, it is also methodologically complex and largely untried. Deep maps are enmeshed in the cartographic metaphor, but instances of deep mapping may extend from spatialized text, to immersive environments, to infinite depth presentation systems, to neogeography, and the geospatial web. The cartographic metaphor is significant because maps and deep maps alike are conceived, designed, and constructed to be interrogated as sources of information, but they are also to be read as graphical texts. Traditional cartographic and geographic information systems (GIS) map products, however, are invariably driven by the pursuit of accuracy where precise measurement, Euclidian geometry, spatial primitives, topology, categorized entities, fields and objects, and appropriate symbolism are combined in some fashion to portray "authentic" representations of reality.

Currently, GIS has become synonymous with mapping in almost every part of the physical, environmental, and human systems. It is a powerful spatial analytical tool, which through its implicit emphasis on space and spatial primitives, spatial data management, and a powerful armory of spatial analytical tools, has contributed significantly to the spatial turn in the social sciences and humanities. Drawing on the backbone of authenticated spatial data, commonly known as spatial data infrastructure, GIS routinely generates cartographic mappings of the physical, surficial, and infrastructural features of the earth: "thin" maps in contrast to "deep" maps. In its ability to construct, deconstruct, and reconstruct spatial entities and regions, GIS has brought about not only a profound technological transformation but an epistemological shift whereby the map has assumed a seemingly unassailable sense of spatiality, surety, and finality.

Despite their seeming neutrality and detachment, maps are far from being objective entities, for they are laden with latent meanings reflecting the predispositions and the intent of their creators and the spaces they are intended to represent. Through necessity, maps are abstractions of reality, for capturing reality in its entirety would involve an improbable 1:1 relationship. As such, maps demonstrate inflections in bias toward inclusion and exclusion and, for this reason, are often deconstructed and reconstructed to reveal and understand intentional or unintentional messages and meanings. The sociotheoretic critiques of the GIS and Society debates of the 1990s, critical GIS, and critical cartography all point to the potential political and social ramifications of GIS usage and raise issues about the privileging of master narratives and categorical causation at the cost of contingent causation and alternative interpretations and renderings of space and place which lies at the heart of deep maps.

At their core, deep maps reflect the ontological and epistemological rebalancing toward the spatial, historical, and social that arose out of the spatial turn. While GIS has been a major contributor to these spatial frameworks because of the intrinsic capacity of maps to discern spatial connections, patterns, and relationships, these spatial scaffolds have also frustrated humanist efforts to understand space and place over time. In this volume, scholars have sought to conceptualize and situate deep maps in a manner that seemingly complement traditional GIS mapping and yet differs from it. Deep maps are intentionally subversive, imprecise, complex, reflexive, sumptuous, and resplendently untidy. They seek to disentangle the multiple realities of human experience lived out in space and place by allowing a heterogeneity of voices to speak in ways that do not privilege one way of knowing over another or force premature reductionism but allow for complexity and for competing outcomes to emerge.

As poststructural representations of place, deep maps contain a plethora of voices, diverse experiences, and worldviews expressed through spatial stories, narratives, and conversations about lives lived out in contingent space and place. They seek to recognize and represent the multiple interests, perspectives, and networks of political power implicit in space and in so doing address issues of structural knowledge distortion and marginalization. They are intentionally reflexive and self-conscious, and by enabling the inherent placeness of human life to emerge, they contain the seeds of their own subversion. Deep maps are the construction and representation of difference, and they acknowledge how diverse forms of knowledge embedded within particular spatial environments are produced and related to daily life and social practice. They are integral to the resurgent interest in space and place which frame and contextualize the spatial narratives and spatial storytelling of these collective and individual experiences. Pathways through deep maps encourage richer considerations of spatial, temporal, and social relationships. Deep maps permit and enable competing perspectives to be considered and through this creative tension provide greater insight and awareness of the role of place and human action in habited space.

Space, place, and spatial thinking are central to the spatial turn and to deep maps. While some disciplines are cognizant of the essential role of geography and space in human behavior, in reality, space is evidenced more as a passive participant, a neutral backdrop, and spatial vacuum to human affairs, rather than as an active participant. Social theory and GIS are odd bedfellows to be motivating and inspiring the spatial turn and not unexpectedly have spawned philosophical and methodological tensions. Unabridged geospatial technologies such as GIS are ostensibly positivist and reductionist epistemologies embedded in the scientific method. Humanist critical discourse is largely text-based and develops narratives of human lives and culture. On many fronts, geospatial technologies would appear to be ill-suited to study the evidentiary materials of humanities scholars whose view of the world is inextricably contingent, nuanced, ambiguous, reflexive, recursive, uncertain, and imprecise. Humanist perspectives are complicated by multiple causal threads and the seemingly unbounded linkages between events and actors. The humanist emphasis is more heavily focused on process, structure, and event in the context of place and place-making than it is on spatial science and GIS-facilitated analyses of spatial relationships and connections. Humanists simulate place through narrative, and yet GIS struggles to handle uncertainty, hybridity, competing outcomes, and variability. The spatial humanities represent a thick weave of contingent behavior, multiple realities, events, and locations that make up the experience of space and place. Understanding place is especially challenging because it is created through human action and is dynamic and constantly being made and remade. As a result, place is more unstable than space. Spatial narrative and spatial storytelling provide a way of reasoning, relaying, and understanding the world through a sequenced ordering and assimilation of events that happen in contextual place.

The intellectual roots of deep maps in eighteenth-century antiquarian approaches to geography, history, people, culture, and place and in the detailed local histories of historical geographers like Hoskins[1] provide a valuable insight into early forms of deep maps. Certainly the deep description of William Least Heat-Moon's *PrairyErth*[2] provides a monumental and finely detailed thick description of a place, Chase County in Kansas, as a textual equivalent to deep maps. In the process, however, *PrairyErth* also demonstrates the limitations of published text and the value of the digital environment in enabling dynamic deep mapping, because the published text fixes a work which is then unable to be augmented, expanded, or be open-ended or reflexive such that additional voices can be heard or for shifting positions to be recorded and evaluated. In their emphasis on intertextuality through hybridization of theater and archaeology, performance artist Mike Pearson[3] and archaeologist Michael Shanks[4] also have provided a powerful image of deep maps as memory practices in which different events and times endure and come together in the material forms of inhabited places where we find the traces and remains of the past. Highlighting the affective ties between people and space and the involvement of people in the places and environments experienced in everyday life, their lifeworlds, also correlates with geographer Yi Fu Tuan's *Topohilia*[5] and the connectedness between human emotion and the physical fabric of landscape as well as to Certeau's[6] spatial stories and practices of everyday life. Equally, the psychogeography of the Situationists International[7] and their positioning of the objective and subjective foundations of knowledge as challenges to the current emphasis on spatial data production and geospatial technology reinforce concern about the emphatic mapping of the material to the exclusion of emotion and experience and the contingent human encounter with the physical, symbolic, and imaginary landscape. Indeed, as Aitken[8] suggests, deep mapping emerges here as redemption for imperious subjectivities and the cartographic spectacle and offers redress through tender mappings that give voice to a language of feelings and memory that connect people to people and places and to storied lives.

Despite, or perhaps because of, these allegories, defining the constituents and characteristics of a deep map is challenging. A deep map is fine-grained, detailed, and situated in multiscalar geographic space which, though bounded, is connected within networks, times, and places. Deep maps are multilayered and draw heavily on multimedia and hybrid methodologies able to capture the many ways of knowing. They represent a weave of narrative and multisourced depictions of people, places, and objects that reflect the cadence and terrain of everyday life all connected through the geographic space that lies behind. Deep maps capture the material and the intangible; the discursive and the ideological dimensions of places; the desires, goals, concerns, and imaginings of people and communities. They also connect the corporeal and immaterial worlds. They are personal, experiential, contingent, reflexive, emotive, and capable of a plurality of spatial pathways through the layered and sedimented strata of information and meaning that have been "buried, exposed, reburied, altered and augmented, diminished and articulated in complex and contingent fashion."[9] They do not claim to be objective or authoritative, but they are more akin to negotiated conversations where contestation and the discourse and dynamics that produce these contestations are an accepted norm. As such, deep maps are open-ended, never final, and are always in a state of becoming. Deep maps are engaging and compelling in representing the renderings and texture of place, both the real and the imagined.

Deep maps are simultaneously a process, a mode of practice, and a product, for in the pursuit of deep mapping, the journey is as important as the destination. The deep map represents a means of conceiving and exploring multiple pathways of discovery through an integrative body of evidence and testimony that enables multiple viewpoints and arguments (whether contradictory, conflicting, or concordant), to be formed, relayed, weighed, and evaluated. Because of their multivocality, deep maps help frame the discussion and the line of questioning and inquiry to be pursued and represent a space for discourse where differing interpretations can emerge. It is here that the reductionist and positivist bent of GIS stands in contrast to the humanist tradition of questioning and of multivalent explanation based on the narrative weave of evidentiary threads. As Tukey[10] would argue, an approximate answer to the right question is worth a good deal more than an exact answer to an approximate question. Thus in practice, deep maps resist reductionism and universal truths and grand narratives in favor of discursive collaboratories abounding with multiple voices, contingency, and contested meanings. As a product, the deep map is integrative, discoverable, searchable, combinable, and capable of reorganization in response to unfolding events, new perspectives, data, and insight: a collage of moments that chart pathways of discovery. The deep map product is not an absolute statement of fact but is transitory and in a state of constant flux and instability and represents a snapshot of that evolving state and the processes and practices that framed the product. The deep map draws on hybrid methods to bridge the gap between the spatial analytical predilection of GIS and the qualitative tropes of social science and the humanities.

Demonstrating the nature of a deep map as product or methodology based on abstract aspirations and without substantive case studies is not easy. Deep maps frame the theoretical but must also situate the pragmatic. They risk creating babel given their combinatorial characteristics to reveal multiple realities; to grapple with inconsistency, incompleteness, and uncertainty in countering universal truths; and to embrace differing methods and ways of knowing. Rather than enhance multiple voices, the deep map may actually drown out any interpretation in a chaos of discord. The mind hates fragmentation and thus it is essential for deep mapping to provide a coherent and sustainable means of organizing disparate evidential records and tracking the pathways of thought and inquiry and engagements through it.

As the previous chapters have illustrated, the examples of deep maps chart a variety of ways in which deep mapping might proceed. GIS is an obvious analog for deep mapping through its layering and emphasis on space as a bounding framework, though the challenge is to shift from a view of humans as entities and data points to an examination of both the material and imaginary world, the official and unofficial, the expert and nonexpert, and the relationships that compose a nuanced, nonreductionist, contingent, and scaled conception of place. Deep maps provide an intellectual landscape within which to create a hybrid methodology that fuses seemingly unrepentant positivist and quantified geospatial technologies with qualitative data and nuanced analysis: a distinct move toward qualitative GIS. Neogeography and the building blocks of Web 2.0 online mapping solutions based on the integrating capability of application programming interfaces provide a potentially exciting framework for deep mapping. Neogeography and the geospatial web provide a configuration of mapping and data solutions that contrast traditional government- and corporate-dominated GIS in augmenting and enriching traditional spatial data with geotagged social media and social networking technologies and multimedia.

Although some scholars view neogeography and its emphasis on volunteered geographic information as a counter to a perceived hegemonic GIS, the geospatial web provides a bottom-up, grassroots, plebian, user-centric, nonexpert, and popular mapping environment that enables people to collate information of their own choice and to create their own maps on their own terms. Through the ability to draw on crowd sourcing, citizen sensors, and mobile location-intelligent social media, the geospatial web enables multiple representations that are highly contextual and personal and that can be mapped and examined and shared among equals. Importantly, these mashups are embedded not just in space but in place and provide for a multiplicity of voices, simultaneity, juxtaposition, subjectivity, and complexity that spans multiple scales. Tapping into these voices provides place-based context and multiple perspectives that represent a radical departure from traditional spatial mapping systems and is far removed from a grand narrative.

The multivocality of neogeography represents a series of conversations that do not attempt to mirror the world but reflect the many fragmented voices, memories, stories and relations about lived place captured in near-real time and that are not subjugated to material geographies. Significantly, neogeography is not dominated by hierarchical scale but is capable of jumping scale through wormholes that connect themes and places and enables multiple pathways of discovery. Neogeographies close the gap between data producers and data consumers such that the user can be both producer and consumer and reflect the personal interactions with place that are tailored to the needs and outlooks of communities of interest: the local collective consciousness that is embedded in both the social and the spatial and portraying communities as they see themselves and wish to be portrayed.

Ghost maps also provide a highly detailed and visual layering of textual, cartographic, and photographic representations that recount the past of a place and the attachment of meanings to a location by making visible the ghostly traces and footprints of past regimes in the landscape. These rich graphical representations of time, change, and events are drawn from the profound complexity of past social life, and places are portrayed through media inscribed with the memories, attachments, and stories of place. The richness and complexity of these inscriptions exceed the capacity of textual narratives to explain or interpret, but ghost maps frame the narratives of place and enable the decoding of the visible and invisible traces of human action in the landscape and successive regimes of power and social ecologies. Each ghost map reveals many stories in a multiple narrative series of people's lives, plots, intrigues, deaths, and dramas in the context of the spaces in which those actions took place.

Geographical text analysis that combines GIS with corpus linguistics assists in developing spatial narratives that reveal how different places are represented over time and identifies events to form a GIS database capable of multiple lines of inquiry and multiple map stories that may correlate, intersect, or depart from each other in time or space. Text reveals complex interpenetrations of the material and the imaginative, the official and the vernacular, and the interweaving of multiple texts reveals how dominant narratives about landscape and place change over time. The analysis of unstructured georeferenced text creates a narrative framework built around a sequential organization of spatial-temporal events where each event is identified by an action verb, actor, location, or time. Connecting these events to form spatial narratives through literary GIS can create geographies of sentiment, emotion, and mood as portrayed through text and place. In a similar vein, ethnographic biographical storytelling, ethnopoetry, and tender mappings represent, as Debord[11] recognized, the weave of personal, memory, emotions, behaviors, and attachment to place in everyday geographies. Emotive and visceral maps represent the nonrepresentable of the personal narrative and of life and the affective relations and transformed connections to other people because emotions irrevocably matter in how humans negotiate events and space. In this way, biography, dialog, narrative, emotion, memory, images, and storied lives are transformed into deep mappings with intersections between lives and places that exist for a moment and then dissolve into dreamscapes. The similarities with Tuan's *Topophilia* and the perceived bond between people and place through the layering of the emotional and spiritual aspects of life to space and place is strong. Deep maps can act as a valuable portal into textual narrative to link the material and the visceral and empower multiple and alternate worldviews where particular identities emerge and change through space and time. Thus tender mappings and evocative writings have a role in deep mapping and the recording of lifeworld moments and movements, and the contested narratives of emotions, poetics, and the political as places become unpacked in terms of the modern subject's spatial and political unconscious.

A number of additional methods indicate the multiplicity and richness of approaches and methods that might be applied to generate and wrestle with deep mapping for no one solution dominates. Qualitative GIS with its focus on spatial multimedia, text, representation, multiple truths, mixed methods, and media-rich digital environments may provide a platform to explore deep mapping and spatial storytelling. When combined with the praxis of participatory GIS and the use of mixed methods in blending authenticated quantitative spatial data with the qualitative knowledge of local communities, the resulting deep map contributes to a greater diversity of opinion and counters the unequal power relations implicit in spatial data creation and use. Neatline and latitude and longitude schematic frameworks create a similar metaphorical loom on which to weave media-rich information of physical and scientific analysis, biography, folklore, narrative, text, memories, stories, and oral histories in the search for emergent spatial narratives. Here, the deep map is enriched by multiple forms of representation that can cast multiple trajectories across maps and scales as spatial stories relay the experiences and relationships that contribute to a deeper recounting of space and place. Layers and topological links within such a deep map database might then refer not just to varied media but to themes, issues, interaction, and experiences that run vein-like through the marbled geographies and mashups of cultural and physical geographies of the imaginary and material. Infinite depth presentation systems provide an organizational framework and a storyboard to navigate through media artifacts of evidence and record and discover latent connections through pathways of alternate readings of the evidence and the differing juxtapositions of the materials. Analog metaphors such as travel guides demonstrate media-rich environments of maps, thick descriptions of history, people and place, cultural heritage, anecdotes, discussions, local personalities, text, graphics, imagery, buildings, and itineraries that ultimately form a deep place-based story. The unfolding spatial story of a place revealed in numerous and inter-related forms are attractive in relating the experience of a place captured through a storied geography.

Thus, deep mapping offers a potentially transforming concept that seeks deeper levels of authenticity and understanding about people and place. Humans experience the world as sensuous, meaningful, reflexive, and highly contingent places. Deep mapping highlights deep contingency and redresses the dominance of meta-narratives by incorporating multiple voices to speak to alternative interpretations. Deep maps incorporate both the tangible and intangible aspects of society and blend quantitative and qualitative data and mixed modes of analysis. The maps reflect the complex interaction of the physical and human environments, and relations and behaviors that are nuanced, nonlinear, branching, and so very difficult to map. Deep maps shift perspective from an emphasis on measured space to that of place and to the connotations, associations, and meanings that people attach to place. In this way, deep maps augment the infrastructural, physical, and environmental themes of conventional mapping to include the cultural, emotional, and perceptual experiences of human behavior. They allow multiple pathways and spatial narratives to be orchestrated in the search for deeper insight, counternarratives, and emergent new topologies that connect people and places. Multiple viewpoints and competing perspectives are not discouraged in deep maps but are accommodated to facilitate open exploration, and the binary disconnect of qualitative and quantitative methods is lessened in pursuit of mixed-mode methods. Ultimately, deep mapping provides a fascinating opportunity to transform seemingly quarantined, objective observations of sterile spatial recordings into an embodied experience of a rich, textured, dynamic, and contextualized place.

NOTES

1. W. G. Hoskins, *The Making of the English Landscape* (Leicester, U.K.: Penguin Books. 1955).
2. W. Heat-Moon, *PrairyErth: A Deep Map* (New York: Mariner Books, 1999).
3. M. Pearson, *In Comes I* (Exeter, U.K.: University of Exeter Press, 2007).
4. M. Pearson and M. Shanks, *Theater/Archaeology* (New York: Routledge, 2001).
5. Y. Tuan, *Topophilia: A Study of Environmental Perception, Attitudes and Values* (New York: Columbia University Press, 1974).
6. M. De Certeau, *The Practice of Everyday Life* (Berkeley: University of California Press, 1984).
7. G. Debord, *Introduction to a Critique of Urban Geography,* reprinted in K. Knabb, ed., *Situationist International: Anthology* (Berkeley, Calif.: Bureau of Public Secrets, 2006).
8. S. Aitken, "Quelling Imperious Urges: Deep Emotional Mappings and the Ethnopoetics of Space," chapter 5, this volume.
9. J. Corrigan, "Genealogies of Emplacement," chapter 3, this volume.
10. J. Tukey, *Exploratory Data Analysis* (Reading, Mass.: Addison-Wesley, 1977).
11. G. Debord, *Introduction to a Critique of Urban Geography.*

CONTRIBUTORS

STUART C. AITKEN is June Burnett Chair of Children and Family Geographies at San Diego State University and Director of the Center for Interdisciplinary Studies of Children and Youth. He is author or editor of eleven books, most recently *The Ethnopoetics of Space and Transformation: Young People's Engagement, Activism and Aesthetics* (2014), and is a founding editor of the interdisciplinary journal, *Children's Geographies*. His research interests include urban and social geography with an emphasis on families and communities, children and youth, and film, as well as qualitative and poststructuralist methods in geography.

DAVID J. BODENHAMER is Executive Director of The Polis Center at Indiana University–Purdue University Indianapolis and Professor of History. He is editor (with John Corrigan and Trevor Harris) of *The Spatial Humanities: GIS and the Future of Humanities Scholarship* (2010). He is the author or editor of ten other books, has written over thirty-five chapters and journal articles, and has made more than seventy presentations on four continents on the application of geospatial technologies to the humanities. He also serves as editor (with Paul Ell) of the *International Journal of Humanities and Arts Computing*.

DAVID COOPER is Senior Lecturer in English Literature at Manchester Metropolitan University (MMU Cheshire). He has published widely on literary geographies and is co-editor of *Poetry & Geography: Space & Place* in *Post-war Poetry* (2013). He has also worked on the exploratory use of digital technologies to map out the literature of space, place, and landscape.

JOHN CORRIGAN is Lucius Moody Bristol Distinguished Professor of Religion and Professor of History at Florida State University. He has authored or edited numerous books on the history of religion, including *Religion and Space in the Atlantic World* (forthcoming). He is editor of the *Chicago History of American Religion* book series, senior editor for American Religion for the *Oxford Research Encyclopedia of Religion*, and co-editor of the *Spatial Humanities* book series.

GRANT DELOZIER is currently a master's student in the Linguistics Department at the University of Texas at Austin. His interests and research are focused toward constructing computational models for the spatial grounding of language.

PHILIP J. ETHINGTON is Professor of History and Political Science and the Co-Director of the Center for Transformative Scholarship at the University of Southern California. His scholarship explores the past as a cartography of time. His recent published works include theoretical works on a spatial theory of history; sociological studies of residential segregation; large-format maps of urban historical change; online interactive Web 2.0 tools, archives, and publications for urban studies; and museum exhibit collaborations. He is currently completing a geohistorical narrative history titled *Ghost Metropolis: Los Angeles since 13,000 BP*.

IAN GREGORY is Professor of Digital Humanities at Lancaster University and uses GIS in the humanities generally and historical research in particular, subjects on which he has published widely. He has written or edited four books, including *Historical GIS: Technologies, Methodologies and Scholarship* (2007, with P. S. Ell), *Troubled Geographies: A Spatial History of Religion and Society in Ireland* (2013, with several other authors), and *Toward Spatial Humanities: Historical GIS and Spatial History* (2014, with A. Geddes), the last two of which are part of the Indiana University Press series The Spatial Humanities.

CONTRIBUTORS

ANDREW HARDIE is Senior Lecturer in Linguistics at Lancaster University. He is a specialist in the methodology of corpus linguistics, with particular interest in corpus design, construction, and annotation; statistical methods, including collocation, as applied to grammar; the languages of Asia; and the application of corpus methods in the social sciences and humanities.

TREVOR M. HARRIS is Eberly Professor of Geography at West Virginia University and is one of the early contributors to the GIS and Society critique of spatial technologies. He researches in the spatial humanities, participatory GIS, the geospatial semantic web, and immersive GIS—a fusion of GIS with virtual reality, serious gaming engines, and other advanced geovisualization technologies.

JOHN MCINTOSH is a GIS analyst with the city of Norman, Oklahoma. His primary research interests are on representation of dynamic geographic phenomena and approaches to support queries and analysis in spatiotemporal datasets.

WORTHY MARTIN is Associate Professor of Computer Science, the Co-Director of the Institute for Advanced Technology in the Humanities (IATH) at the University of Virginia, and the Associate Chair of the Department of Computer Science. IATH explores and develops information technology as a tool for scholarly humanities research. He also served as advisor to the National Initiative for a Networked Cultural Heritage. His primary research interest is dynamic scene analysis, that is, computer vision in the context of time-varying imagery, and image databases.

PAUL RAYSON is Senior Lecturer in Computer Science and the Director of the UCREL interdisciplinary research center at Lancaster University and has published widely on natural language processing and corpus linguistics methodology. His applied research is in the areas of online child protection, learner dictionaries, and text mining of historical corpora and annual financial reports.

NOBUKO TOYOSAWA is a postdoctoral fellow in Early Modern Japanese studies in the Department of East Asian Languages and Civilizations at the University of Chicago. She is currently completing a book titled *Placing Japan: National Imaginaries and the Formation of Historical Knowledge in the Tokugawa and Meiji Eras*.

BARNEY WARF is a Professor of Geography at the University of Kansas. His research emphasizes producer services and telecommunications, including fiber optics, the satellite industry, offshore banking, international producer services, and the geographies of the internet. He has also written on military spending, voting technologies, the U.S. Electoral College, and religious diversity. He has authored, co-authored, or co-edited seven books, two encyclopedias, thirty-two book chapters, and roughly a hundred refereed journal articles. He currently serves as editor of *The Professional Geographer* and co-editor of *Growth and Change*.

MAY YUAN is Ashbel Smith Professor of Geospatial Information Science in the School of Economic, Political, and Policy Sciences at University of Texas at Dallas. She is the associate editor (Americas) of the *International Journal of Geographic Information Science* and is a co-author of *Visualization and Computation of Dynamics in Geographic Domains* (2007). Her research centers on geographic dynamics and spatiotemporal representation of physical, social, and recently cultural processes from a wide range of data sources, including texts.

INDEX

actants. *See* networks
Adolphus, Svenja, 170
agency: human, 20, 59, 62, 203, 211, 221; of space and time, 23
analysis: digital, 45, 64, 150, 159, 173, 188–190, 198–200, 213; semantic, 160–161; spatial, 59, 166–167, 198–200, 205–206; spatiotemporal, 183, 199; spatiotextual, 19, 173–175; textual, 18–19, 64, 72–87, 150–153, 167, 173, 182–184, 191, 207–208, 230
Antarctica, 34–35
anti-imperialism, 116–117, 227
archaeology, 19, 36, 44, 211–218

Bakhtin, Mikhail, 13
Bayes method (classification), 187–188
Benjamin, Walter, 60, 102–104
Blake, William, 17
Bodenhamer, David J., 103, 172
Bourdieu, Pierre, 61–62
Brión, Spain, 143–146
Bruno, Guiliana. *See* tender mappings
Buddhism, 81–83, 86–87

capitalism, 31, 104, 142
Casey, Edward, 14, 16, 169
Cash, Johnny, 49–51
Certeau, Michel de, 31, 42–43, 54, 60–62, 65, 170, 226

Chaco Canyon Research Archive, 211–218
chronotope, 13
Civil War (United States), 181–184, 194, 204–207
Clay, Jenny Straus, 218–219
Coleridge, Samuel Taylor, 167
collocation, 160–166, 170
commemoration, 19, 82–83, 117–118, 145, 211–212
concordance, 154–155
Constituent Likelihood Automatic Word-tagging System (CLAWS), 156
corpus linguistics, 152
crowd sourcing, 29, 49–51, 138, 229

Debord, Guy, 31–32, 102, 118
Deconstructionism, 14, 60, 224
deep geography, 28–29, 135
deep maps, 1–2, 20–23, 38–39, 65, 72, 89–96, 103, 143–146, 203, 223–232; epistemics of, 31, 135–137, 172, 224–227
density smoothing, 163, 165–166
Dewey, John, 140
Dyer's Compendium (Frederick Dyers), 181–184

Ekiken, Kaibara, 75–79
emotion, 21, 32, 104, 108–110, 167–168, 230
empire, 80–84, 91–92, 104, 107
Entrikin, J. Nicholas, 58

239

environment, 8–10; digital, 45; experience structured by, 19, 21, 33, 36, 57, 134, 147, 207, 230–231; natural, 74; urban, 32, 43, 65, 90–96; virtual, 19, 45–46. *See also* experience: in context
Ethington, Philip, 90–96; ghost maps, 230
ethnography, 29, 35, 44–45, 113–115, 118–130, 145, 172. *See also* narrative: and historical writing
ethnopoetics, 104, 111–112
Euclidean space, 32, 57, 189
experience, 2, 13, 32; collective, 112; in context, 18–19, 36, 42, 61, 74, 88, 110, 118, 167, 190, 232; lived, 17, 36, 42, 59–62, 89, 108, 140–141, 224, 226; and narrative, 13–14; of place, 16, 32–33, 44, 56, 114–115; of space, 10, 16, 22, 116–117, 147, 190

femininity, 109–110
folklore, 34, 35
Foucault, Michel, 55, 61–62, 72, 136
frequency profiles, 159
Freud, Sigmund, 8, 106
fudoki, 75–77

Gaddis, John Lewis, 17, 179
gaming, 19–20, 42
gazetteers, 157, 186–189
gender, 143; femininity, 109–110; masculinity, 106; patriarchy, 106
genealogies of emplacement, 63–67
genealogy, 63
geographic information systems (GIS), 1, 10, 18, 29, 145, 157, 183, 198–199; and archaeology, 19; epistemics of, 2, 10, 23, 135, 223, 228; and narrative, 180, 191–199; participatory GIS (PGIS), 29, 45; purposes of, 20–23, 29, 137, 152, 167–168, 179–180, 224
geography: academic discipline, 15, 60, 137–138, 173; and history, 15, 35, 73. *See also* neogeography
geonarratives, 41–42
geopower, 103, 111, 116
geovisualization, 18, 45

ghost maps, 90–96, 230
globalization, 2, 90–92, 134, 143, 146
Google: Earth, 35, 46, 79, 163; Maps, 35, 46, 138, 163; Streetscene, 45
grammar. *See* parts of speech
Gray, Thomas, 167
Gregory, Derek, 59–60
Grosz, Elizabeth. *See* geopower

Habermas, Jürgen, 140–141, 147
Habitus, 61–62
Hartshorne, Richard, 57–58, 60
Harvey, David, 18, 58–59
Heat-Moon, William Least, 135, 226
Heidegger, Martin, 14, 55–57
history: academic discipline, 5, 10, 15, 29, 35, 225; and geography, 15, 35, 73; local, 44, 80, 84–85, 90, 135–136, 144; and myth, 12, 76–77, 218; narrative, 12, 63, 88, 97, 185; national, 77–78; oral, 39, 42, 113, 145; spatial, 9, 41, 72, 168, 172, 179. *See also* mapping: historical; time: historical
Homer, 218–221
humanities, 10–12, 18, 72, 134, 150–151, 172; spatial, 20, 223–225

Iliad, 218–221
intertextuality, 35, 108–110, 168–169

Jacobites, 105
James, William, 140, 147
Japan, 75–89, 97
Javascript, 137
Johnny Cash Project, 49–51
Joyce, James, 74–75

Keokuk, Iowa, 54, 67

Lake District, England, 153, 163; Mapping the Lakes Project, 167–173
landscape, 35, 75, 84, 88, 97, 170; historical, 92; mythological, 221; Three Landscapes Project, 36–37
Latour, Bruno, 61–62
light detection and ranging (LiDAR), 19

INDEX

Lincoln, Abraham, 194
London, England, 43, 153
Los Angeles, California, 90–96

mapping: Cartesian, 18, 114, 139; cartographic, 17, 29–30, 75, 167, 223; conceptual, 17, 73; historical, 17, 72, 92, 169, 179, 196, 205–206; software, 138; textual, 74–77. *See also* deep maps; tender mappings
Mapping the Lakes Project, 167–173
Marxism and Marxists, 31–33, 58–59, 102–103, 226
masculinity, 106
Massachusetts Bay (Salem Witch Trials Project), 207–211
Massey, Doreen, 15, 20, 58, 59
McLucas, Clifford, 37–39, 136
memory, 10, 14; cultural, 22, 36, 39, 72, 82–83; and emotion, 14, 113, 117, 119; of places, 44–45, 103, 145. *See also* commemoration
metaphor, 18–19, 29, 33–34, 50, 73, 139, 223, 232
Moretti, Franco, 151, 174
myth(ology), 12, 76–77, 85–88, 218–221

named entity recognition (NER), 187
narrative, 2–3, 11–13; and historical writing, 12, 17, 20, 22, 97, 179; meta-narratives, 41, 49–50, 232; and space, 20, 28, 39–41, 65–66, 84–86, 146, 168, 180. See also history; spatial narratives
National Park Service (U.S.), 212
natural disasters, 76
natural language processing (NLP), 152
Natural Language Toolkit (NLTK), 184
neogeography, 134–139; politics of, 138, 142–143, 238
networks, 21, 61–62, 90–92, 145, 151–152, 180, 206
New Historicism, 20

optical character recognition (OCR) software, 153
Ovid, 90

Paris, France, 110
parts of speech, 74, 156, 159, 184–190, 199
patriarchy, 106
Pearson, Mike and Michael Shanks, 31, 35–36, 103, 135, 226
performativity, 35–36, 141
phenomenology, 56
place, 7, 15–16, 29, 61, 168; place-making, 29, 55, 63, 67, 72, 169; place-names, 76, 92, 158–160, 168–170, 188. *See also under* experience; space
placialization, 14, 16, 169
pilgrimage, 86
postmodernism, 5, 14
poststructuralism, 136–141, 224
pragmatism, 140
Prezi, 46, 48, 66–67
print culture, 78
public sphere, 102, 140
Pueblo Bonito, 213–218

Quasha, George, 112

railroads, 206
Rorty, Richard, 141
Rothenberg, Jason, 112–113

Salem Witch Trials Project, 207–211
scale, 17, 23, 175
scale shifting, 179–180
Shakespeare, William, 44
Shanks, Michael. *See* Pearson, Mike and Michael Shanks
Situationist International, 31–33, 102, 226
slaves, 92, 191–198
space, 2, 8–9, 107, 180, 185–186; imagined, 75; metaphorical, 18; and place, 7, 14–16, 22, 43, 62–63, 232; sacred, 87–88. *See also under* experience; narrative; time
Spanish Empire, 91–92
spatial narratives, 12–14, 65–66, 96, 163, 199, 204–205, 221. *See also* narrative: and space
spatial turn, 1, 7–10, 20, 64–66, 134, 223–225
spatiotextual analysis, 19, 173–175

story, as different from narrative, 40–42. *See also* narrative
storytelling, 40–42, 50, 226

technology, and experience of space and time, 7, 106–107, 144
tender mappings (*carte du pays de tender*), 103–104, 107–110, 113, 116–117
text, 64, 89, 150, 173–174, 226; and place, 89. *See also* analysis: textual; intertextuality
text-mining, 151, 156, 174
thick description, 35, 172. *See also* ethnography
"thin" maps, 29–31, 223
Three Landscapes Project, 36–37
Tilley, Christopher, 22
time, 2; ancient, 77, 80, 87, 95–97; historical, 57, 88–89, 199; in maps, 72, 198–199, 209, 218–219; and space, 13, 134, 180, 198, 203, 210–211, 231

topology, 46, 67–68, 135, 223
tourism, 43–44, 232
Tuan, Yi Fu, 15, 21, 31, 56–57, 226
Turner, Frederick Jackson, 8
Twain, Mark, and rules for storytelling, 40

United States Geological Survey, 92
urban environments. *See* environment: urban
Uto-Aztecans, 92–93

virtual environments, 19
visualization, 18, 45, 84, 220

Web 2.0, 46–48, 134, 138, 144–146, 150, 228
Westphal, Bertrand, 168–169, 171
Williams, Raymond, 22
Wordsworth, William, 162, 171

XML, 137, 154